忍術及手裏剣

Ninjutsu and Shuriken

By Eric Shahan

ISBN: 978-1-950959-89-1
Imprint: ERIC MICHAEL SHAHAN

1. *Negishi School Shuriken*
   Kawauchi Tetsusaburo
   川内鉄三郎 (1876~1943)
   1935

2. *Research into Shuriken*
   Naruse Kanji
   成瀬関次 (1888 ~ 1948)
   1941

3. *The Japanese Throwing Arrow*
   Murakawa Kiyoshi
   村河清 (1882~1913)
   1928

4. *The Japanese Throwing Arrow*
   Kitamura Akitaro
   北村明太郎 (?~1933)
   1926

5. *Ninja Tools*
   Fukuyama Matsui
   福山松翠 (?~1967)
   1953

6. *Secrets of Ninjutsu*
   1776
   Author Unknown

7. *Secrets of Moen School Shuriken*
   Edo Era
   Author Unknown

Negishi School Shuriken

Kawauchi Tetsusaburo
川内鉄三郎 (1876~1943)

Excerpt From:
Nihon Budo Ryusoden 日本武道流祖伝
Legends of the Founders of Japanese Martial Traditions
Published 1935

Negishi Ryu Shuriken

Founded by Negishi Shorei 1833~1897

Negishi Shorei was a retainer of the Annaka Domain in Joshu in what is today Gunma Prefecture. From a young age he studied the Hokushin Itto School of sword under legendary sword master Chiba Shusaku as well as his son. In addition, Negishi later studied under Shusaku's senior disciple Kaiho Hanpei (1822~1863.) After mastering the inner mysteries of Hokushin swordsmanship, Negishi was appointed head of Kaiho's Dojo.

Later, Negishi embarked on a martial arts pilgrimage, travelling throughout north-eastern Japan and training at every Dojo in the area. Upon returning home he was appointed sword instructor to the Samurai of Annaka domain. His fame as a martial arts instructor spread and he was nicknamed "The Little Tengu (mountain goblin) of Joshu (Gunma Prefecture.)"

Shorei studied Shuriken from a man named Kaiho Hanpei and he later transmitted his teachings to Tonegawa Magoroku, a retainer of Tatebayashi Domain in what is now Gunma Prefecture. Magoroku was an expert in the sword arts in addition to spear fighting and the equestrian arts. Taking what he learned from Shorei, he continued to refine the Shuriken techniques he had been taught until he had perfected his own style.

Origin of Shuriken Jutsu

The origins of the art of throwing Shuriken can be found in *Kyujutsu*, Japanese Archery. Many schools of archery have passed down traditions regarding how to make and use two types of weapons not shot from a bow: *Tezuki-ya,* Hand-thrust Arrows, and *Uchine,* Throwing Arrows.

These can be used both as projectile weapons to stop an approaching enemy Samurai or as a close-quarters blade to stab a Samurai when he gets into range. Just as bladed weapons for stabbing emerged, so in turn, did these weapons give rise to the practice of throwing short blades at an opponent. This came to be called *Shuriken ni Utsu*, "to strike an opponent with a blade the size of the palm of your hand."

Translator's Note: Hazu Yari

Hazu Yari 弭槍 End of the bow spear

After an archer runs out of arrows or his last bowstring breaks, he can defend himself using the spearpoint attached to the top end of his bow.

The *Zohyo Monogatari* 雑兵物語 *The Foot Soldier's Story* (attributed to Matsudaira Nobuoki (1630~1691,) a Daimyo and advisor to the Shogun) is a how-to guide for low-level Ashigaru "light foot" Samurai.

Ashigaru included spearmen, archers, matchlock squads, grooms and porters who made up the bulk of seventeenth-century armies. *The Foot Soldier's Story* consists of thirty first-person accounts by Ashigaru and served as a manual for new recruits. The book contains several references to soldiers using *Hazu Yari* during the Muromachi Era (1336 ~ 1573.)

Thus, *Shuriken Jutsu*, the art of making and throwing small blades, was gradually developed and refined over time and came to be valued as its own unique and important branch of martial arts. Shuriken were likely derived from a combination of:

- Attaching a spearhead to the end of a bow (*Hazu Yari*)
- Crafting arrows specifically designed for throwing (*Tezuki-ya*)
- Throwing short swords or knives (*Uchimono*)

Interestingly, as Shuriken techniques developed, swordsmen devised new defensive methods in response.

The earliest reference to *Tezuki-ya*, Hand-thrust Arrows, can be found in the *Chronicle of Great Peace* from the 14th century.

As for the earliest reference to Shuriken throwing, that can be found in *Various Tales Recorded by Jozan,* a collection of four-hundred and seventy interesting stories about 16th century Samurai warlords.[1] There are episodes describing Shuriken being employed in combat such as,

> Tsuto Jinsuke was pursuing Suganuma Gyobu. Tsuto shot arrows at Suganuma but missed. Drawing his blade Tsuto threw it like a Shuriken, striking a glancing blow on Suganuma's head. So the story says.

And,

> Yoshihiro fashioned one with a black leather handle. He wore a helmet adorned with bear fur. Yoshihiro, on horseback and armed with a long sword, engaged another mounted rider named Inoue. Though Yoshihiro was knocked from his horse, he drew his Wakizashi and threw it like a Shuriken, striking Inoue in the thigh on his bow-hand side. In that moment, another Samurai, Oguri, closed in and began grappling with Yoshihiro, eventually taking his head. So the story says.

---

[1] *Jozan Kidan* 常山紀談 by Ishikawa Jozan 石川丈山 (1583~1672)

Translator's Note: *Tezuki-ya*, Hand-thrust Arrows

Illustration of Takainba Zensho

Translator's Note: *Tezuki-ya*, Hand-thrust Arrows

Taiheiki: Chronicle of Great Peace
Volume Fifteen: The battle of February 28th, 1335

Takainba Zensho was armed with *Tezuki-ya*, Hand-thrust Arrows, a weapon that did not require a bow. Zensho stood like a giant guardian statue in front of a temple constructed at the front line of the fighting and proclaimed,

> It was I who led the most cunning advances last year at the Battle of Three Wells Temple! I came from Myokan in Echigo Domain! My name is Takainba Zensho! You up there in the castle, which of you will fall to my arrow! Let's play!

With that he hurled one of his arrows at a *Sama*, or a gap in the castle walls for archers to shoot through. His arrow flew true and struck an armored Samurai in the *Sendan no Ita*, a flap that serves to protect the armpit on the right side. It went straight through that piece and out the back plate. It pierced both pieces with such force that six centimeters of the arrowhead were sticking out the man's back. The soldier, unable to utter a word, fell from the tower dead.

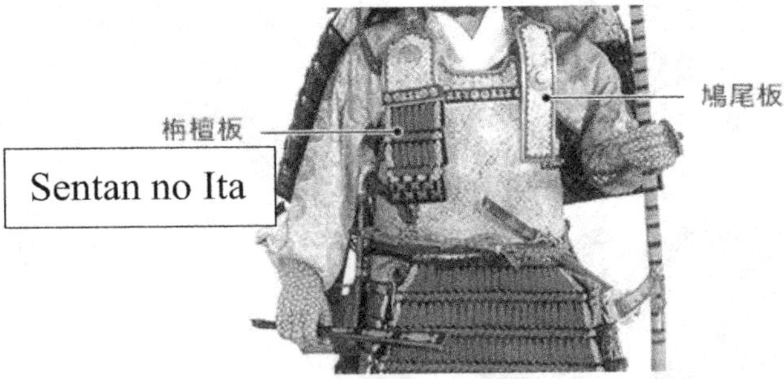

The other enemy soldiers, having witnessed this attack exclaimed, *What monstrous power! Surely nothing a mortal man can do!*

Inspired by the fearful air that was spreading amongst the defenders, the thousand Samurai surrounding the castle suddenly drew their sword in unison and attacked.

Thus we have examples of Samurai using short swords or knives as Shuriken starting from the middle of the warring states period (1540~.) While initially there were only one or two techniques that were employed in extreme situations, Samurai continued to develop these techniques and the art began to flourish, eventually coalescing into refined techniques, which were then passed down to the following generations.

In the *Osaka Gunki,* War Record of the Osaka Campaign, which records stories of the battles that took place in the early 1600s, there is an incident regarding Shuriken.

Ogasawara Tadamasa was fighting an enemy Samurai armed with a spear. The Samurai had found a gap in Ogasawara's breastplate with his spearpoint and was readying to impale his foe. However, as Ogasawara was skilled in Shuriken, he quickly drew a short blade and threw it, causing his attacker recoil, pulling his spear out in the process. In the end, Ogasawara survived with only a shallow wound.

Also, a passage from *Various Tales from a Refined Recluse* records that a man named Takemura Musashi was highly skilled in Shuriken.[2] He could hit a peach floating down a river with a fifteen-inch blade.

From this we can see that Wakizashi short swords and daggers were often thrown as Shuriken. Eventually, the technique of throwing knives or short swords began to be referred to as *Uchimono*, striking implements.

---

[2] *Zenan Zatsuwa* 善庵雑話 by Asakawa Zenan 朝川善庵 1850.

The *Yagyu-ryu Shin Hisho*, Yagyu School New Secret Teachings, written by Sano Katsufuru in 1716 contains a section called *Uchimono*. It says,

> When facing an opponent who is wielding two swords, he will likely be holding his short sword in his left hand. He will throw that blade in order to distract you and disrupt your initial strategy. If facing such an opponent, respond by defending with a cross-shaped block when he throws with his left.[3]

Also,

> If you intend to cut down the incoming strike, waiting until after your opponent throws will mean you will be too late. Instead, you must focus all your attention on your opponent's fist.
>
> Regardless of the distance, near or far, focus on linking the movement of your own fists with that of the hand your opponent will throw with. If you cut while matching your opponent's motion with your own hands, then you can cut the thrown weapon down in midair with a *Juji*, cross-shaped cut. This state of mind is called, "Seeing the Shuriken and striking the Shuriken."

---

[3] A "cross-shape" is when you knock an attack down with your sword perpendicular to your opponent's attack. In this case, you are knocking down the sword your opponent throws.

Translator's Note: *Kozuka* and *Kogai*

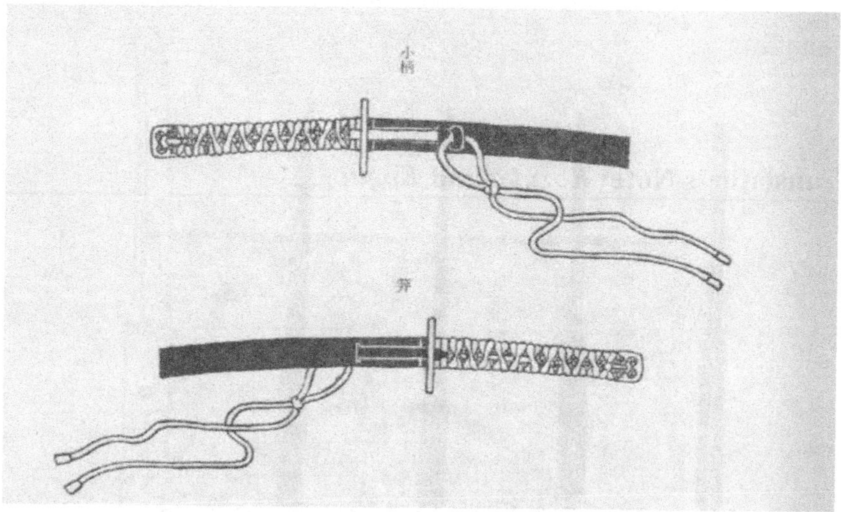

Top: *Kozuka*, a small knife housed on the outside of the scabbard.
Bottom: *Kogai* hairpin housed on the inside of the Katana scabbard.

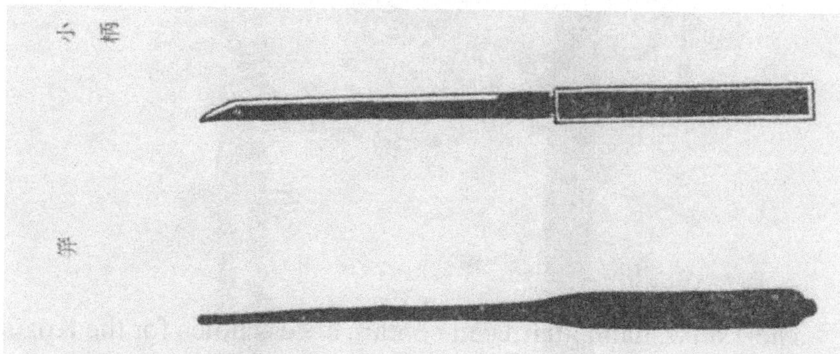

Top: *Kozuka* 小柄 was a utility knife for things like cutting pieces of scrap paper for writing a note.
Bottom: *Kogai* 笄 Is a hairpin kept on one side of the sword. Though it is categorized as "sword equipment," it was generally a tool to groom yourself. It had various uses, such as scratching spots around your topknot, removing earwax, fixing messy hair, and so on. Some schools also teach it can be used as a weapon.

Translator's Note: ***Kozuka*** **and** ***Kogai***

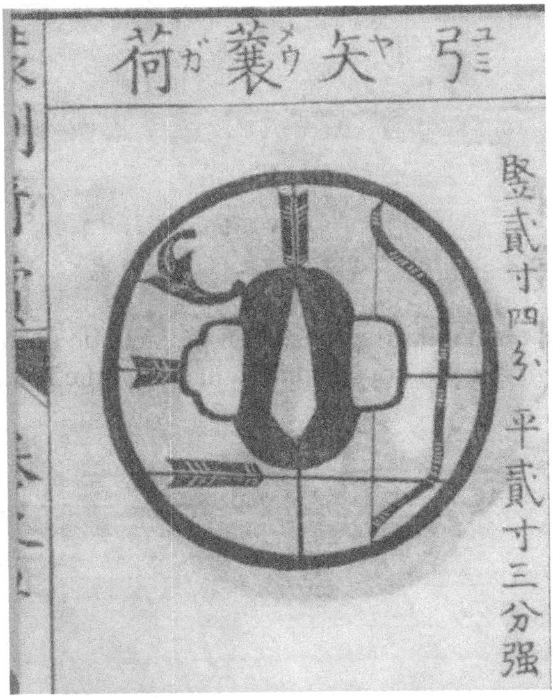

The *Tsuba*, hand guard can contain access holes for the Kozuka and Kogai. This Edo Era hand guard, featuring a bow and arrow motif, has the Kogai hole on the left and the Kozuka hole on the right.

Chokuda/ Jikida 直打 Straight Throw
Kaitenda/Kaiten Uchi 廻轉打 Rotating Throw

In the *Kobunko* encyclopedia, Professor Mozume writes that the greatest master of Shuriken in the land was the military tactician Yui Shosetsu.[4]

Further, Dr. Tsubouchi's writes in the *Collection of Early Modern Records*, the story of how Yui Shosetsu had an intense Iai, sword drawing duel, against Sekiguchi Hayato, the head sword instructor to the Kii domain, what is now Wakayama Prefecture and Mie Prefecture.

> During the duel Shosetsu appeared to pause for a moment before leaping back and drawing his Kogai hairpin (not his Kozuka) from his scabbard and throwing it as a Shuriken, striking the hem of Hayato's Hakama and pinning it to the floor, thereby achieving a win.[5]

---

[4] Mozume Takami 物集高見（1847~1928）was a *Kokugaku* "Study of the Japanese Nation" scholar. This was a movement to refocus Japanese scholarship away from the study of Chinese, Confucian, and Buddhist texts in favor of early Japanese classics.

Yui Shosetsu 由井正雪 (1605~1651) was a Japanese military scholar.

[5] Tsubouchi Shoyo 坪内逍遥 (1859 ~ 1935) was a Japanese author, critic, playwright, translator, editor, educator, and professor at Waseda University.

Another story,

> There was a monk called Kyukaizan from Tenjintani at Zojoji Temple.[6] He was the son of Akashi Takenori, formerly of the Osaka side,[7] and in childhood went by the name Kikumaru.[8]
>
> At the age of thirty-six he was a man of unparalleled strength, and was especially adept at the martial art of throwing Shuriken. Later he left the temple and returned to lay life, taking the name Yoshida Hatsu-Uemon, and moved to Meguro Gyonin-zaka.

It is said that Yoshida later became a retainer under Shosetsu and participated in the Keian uprising, serving under the general leading the defense of Osaka.[9] However he was surrounded while staying at Arima Onsen and captured. According to the previously mentioned *Collection of Early Modern Records* Arima said,

> I wish to avoid any unnecessary bloodshed, but allow me to show you what I can do. With that, Yoshida took a dozen arrows from inside his shirt and, aiming for the pine needles hanging off a tree in the garden, threw the arrows like Shuriken so the pine needles dropped off one by one.

Clearly Yoshida was demonstrating *Uchine Jutsu*, Throwing Arrow Techniques. His method of *Uchimono*, throwing weapons, was a *Chokuda Ho*, throwing on a horizontal line directly at the target.

---

[6] Zojo-ji Temple was one of the largest temples in Edo.

[7] During the 1614~1615 Siege of Osaka, Akashi fought against Tokugawa Ieyasu's forces from inside Osaka Castle.

[8] Akashi Takenori 明石全登 (1566~1618)

[9] The Keian Uprising 慶安事件 was a failed coup attempt against the Tokugawa shogunate in 1651 by disgruntled Ronin. The leaders were Yui Shosetsu (who later committed Seppuku) and Marubashi Chuya (who was crucified and his family beheaded.)

In the Kanei Era there was a master Shuriken practitioner who lived in Osaka by the name of Mori Gentatsu.

Gentatsu owned a lumberyard but one day a stack of boards fell on his foot, causing him to become lame in one leg. Furious at this injury he began developing a method for throwing five inch nails into wooden targets. Eventually, after long practice and refinement Gentatsu developed his own school.

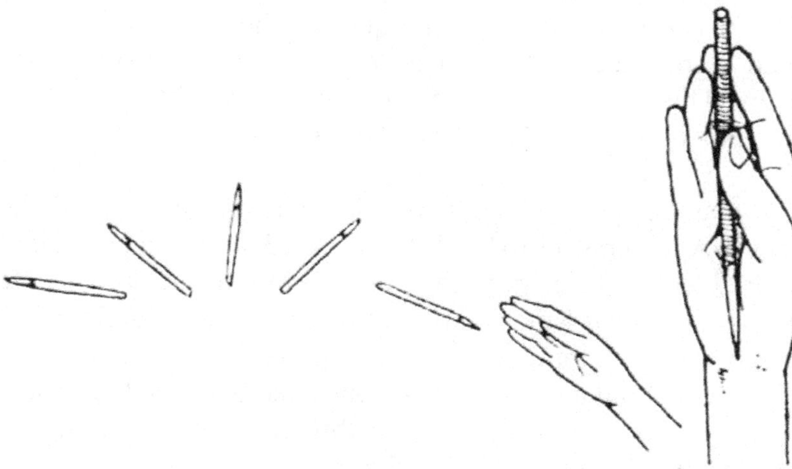

The method was one where a tip of the nail was held with the point was facing the base of the palm when thrown. This method of throwing a Shuriken so that it rotates in midair before striking represents the dawn of two styles of throwing: Straight Throw and Rotating Throw.

Gentatsu's methods for throwing and defending against Shuriken have been transmitted down to the present.

The distant founder of Straight Throw: Kadono Hirohide

In a book called *A Guide to Japanese Readings of Chinese Words*[10] from around the Meiwa~Anei eras 1764~1781, contains an entry for the word Shuriken:

> The word Shuriken しゅりけん is written as 手裡剣 in Kanji. The Kanji are Hand + Reverse (of the hand, palm) + Sword. It is used in Gekken no Jutsu, sword fighting techniques.

Around this time, a Samurai from Sendai Domain named Kadono Hirohide of Sendai devised a method of throwing needles that was based on a combination of *Tanto Chokuda*, Straight Knife Throw and *Uchine*, Throwing Arrow, techniques.[11] The throwing darts he developed were about three inches long and had a heavy, cone shaped tip. This may be regarded as the origin of Negishi Ryu Shuriken. In a book titled *Tales of Oshu*[12] it states:

> A man called Kadono Hirohide, who served around the Meiwa~Anei eras (1764~1781) received a stipend of eight hundred Koku of rice per year from the lord of the domain. He was a skilled swordsman and was particularly adept at Shuriken as he had spent a long time developing his technique.
>
> He could hold a single needle concealed between his middle finger and the next finger. When thrown, this needle-shaped Shuriken would never fail to strike.

The underlying concept of this weapon was supposing that if you were faced with an enemy, you could blind both his eyes. If you were able to do this then you would have no need to fear an opponent no matter how fearsome he is.

---

[10] *Wakun no Shiori* 倭訓栞 Compiled by Tanikawa Kotosuga 谷川士清 (1709~1776) and published in 1830

[11] Kadono Hirohide (?~?) was a swordsman in the mid-Edo period.

[12] *Oshu Banashi* 奥州波奈志 By Tadano Makuzu 只野真葛 (1763~1825,) daughter of an Edo physician, who wrote commentaries on Japan's social and political problems.

Also,

> When demonstrating before the head of the domain at the Old Shiba Detached Palace, Kadono asked his lord to name a target. The Daimyo indicated he wanted Kadono to strike the 'claws of a four-legged creature' that decorated a cedar door panel. Kadono threw twice, both times hitting the spot indicated, much to the amazement of the elder lord.

It is also recorded that the marks where the Shuriken struck remained clearly visible up until a 1724 fire destroyed the building

Kadono's method was to approach an enemy and swiftly throw at both eyes. He called it *Fusatsu-Sakuto,* ending a fight with a non-lethal strike.[13] This method spread to many domains in the area that is now Iwate Prefecture.

Negishi Ryu

Kaibo Hanpei, teacher of Negishi Shorei, was the sword instructor for the Samurai of Annaka Domain in Joshu. Kaibo was a senior disciple of Chiba Shusaku and was also versed in many martial arts. In particular, he was a skilled Shuriken practitioner. Later, Kaibo was summoned to serve the ninth lord of Mito Domain Lord Tokugawa Nariaki. It is said that the lord ordered him to aim for a pigeon that was resting in the garden. Kaibo drew a Shuriken and struck the bird down, yet when the Shuriken was removed, the pigeon suddenly began flapping and flew away.

Shorei, having served as Hanpei's assistant instructor, further improved upon the needle devised by Kadono and established the foundations of what later became Negishi School of Shuriken. This refinement was completed by Tonegawa Magoroku. The name "Negishi School" was given later by disciples in admiration of the founder.

The Negishi School traditions have been handed down over the years and they continue to be taught in the present through Naruse

---

[13] *Fusatsu* 不殺 non-lethal
Sakuto 削闘 bring to an abrupt end

Kanji. (Three generations before Naruse Kanji there was a man named Masanori, who served as a retainer of the Matsudaira Family in Kuwana. Masanori was the Iai, sword-drawing, instructor for the domain and taught Yamamoto School sword techniques. The Yamamoto School is derived from the Sekiguchi School of sword. It is said that the Shuriken throwing Kata and defense against Shuriken Kata from the Yamamoto School are still transmitted today.)[14]

The spirit of Negishi School Shuriken lies in the fact the techniques represent the *Ura*, reverse, of sword techniques. The sword is a weapon that when employed results in death. On the other hand, the blade of a Shuriken does not kill. The Shuriken is a non-killing sword, while a Katana is the life-taking sword. The sword kills, while the Shuriken allows you to control an opponent by bringing a duel to an abrupt end in a non-lethal fashion.

Although the effective striking range of a Shuriken is 5~6 Ken, 30~35 feet, employing a Shuriken at that distance is not aligned with the principles of swordsmanship, so you would typically not throw until your opponent had closed to within 3 Ken, 18 feet. As Shuriken techniques are part of sword fighting it is not treated as a projectile weapon, like bows or matchlocks. Thus it is forbidden for practitioners attempt long-distance strikes which would be relying more on luck than skill.

The fundamentals of the Negishi School are taught in three Kata, which progress from easy to difficult and from slow to fast. Though the fundamental teachings only comprise three techniques, there are a multitude of variations within them.

There are four *Sento Kata*, Fighting Techniques, the first and second of which are used in one-on-one situations. These techniques focus on changing between stillness and movement. The third technique is for when you are facing enemies both in front and behind. It uses movements similar to the *Zengo Uchi*, Striking Forward and Back, cuts found in Iai, Sword Drawing. The fourth Fighting Technique is when you are in a life-or-death situation, and must rely on a complex combination of all four *Sento Kata*, while taking into account what weapon your opponent is armed with.

---

[14] Brackets by the author

Research into Shuriken

Naruse Kanji
成瀬関次 (1888 ~ 1948)

Excerpt From:
Bugu Kacchu no Kenkyu 武具甲冑之研究
A Study of Arms and Armor
Published 1941

Research into Shuriken

The *Record of the Essence of Negishi School Shuriken Techniques*,[15] contains the following passage,

> Shuriken Jutsu is an art derived from *Uchine*, Throwing Arrow, techniques found in Japanese Archery combined with *Uchimono*, Throwing Knives, found in Kenjutsu. These two methods, having been removed from their original art have been fused into a single discipline.
>
> Therefore, the throwing stance used in Shuriken resembles that found in archery, while the mentality is from sword fighting. Throwing Shuriken has nothing to do with using the fingertips. Its subtle application arises from taking a stance that is neither rigidly upright nor bent, that alternates freely from Yin to Yang, and thereby enabling you to control an opponent armed with a great three-shaku long sword.[16]

This passage clearly outlines both the history and the essential principles of Shuriken Jutsu.

When the *Koji Ruien* encyclopedia was being drafted, experts debated how to classify Shuriken Techniques.[17] Although outwardly the throwing stances resembles ones found in archery, its true nature is clearly aligned with the spirit of swordsmanship, and thus it was grouped together with Iai, sword drawing techniques, in the section on sword arts. It is interesting to note that definition they arrived at closely matches the above passage.

---

[15] *Negishi-ryu Shurikenjutsu Yoroku* 根岸流手裏剣術要録 Record of the Essence of Negishi School Shuriken Techniques. No further information about this book.

[16] Three feet in length. However this is describing the length of the blade, not the entire sword. The phrase “Three Shaku Sword” is often used to describe a large sword, not necessary its actual length.

[17] The *Koji Ruien* 古事類苑 is a Japanese encyclopedia commissioned by the Meiji government. From 1896 to 1914 over a thousand volumes were compiled.

Uchine Jutsu Throwing Arrow Techniques

If you wanted to take a historical look at the development of Shuriken, the oldest references appear in the *Collection of Ten Thousand Leaves* (759 AD) and, much later, in the *Chronicle of Great Peace* (1300s.)[18] One poem in the *Collection of Ten Thousand Leaves* contains the line,

…投ぐる箭の遠さかり居て
…the thrown arrow flies far

Thus, we can see that it is likely, even in this early era, that *Nage Ya,* Throwing Arrows, were already being employed. These Throwing Arrows evolved into what were called *Tezuki-ya,* Hand-thrust Arrows. Later this name evolved into *Uchine,* Throwing Arrows. These were weapons designed for close-quarters combat. If an opponent was close, the weapon could be thrust at the enemy like a spear and when somewhat farther away, they could be thrown.

The *Chronicle of Great Peace* contains the story of Inba Zenmura who carried a great mountain of thirty-six large arrows on his back and hurled them by hand, piercing Samurai straight through their armor.

The *Biographies of Notable Masters of Japanese Martial Arts,* contains the following passage,

> The *Tezuki-ya*, Hand-thrust Arrow, differs from Shuriken as they have feathers attached to the end. Therefore, no matter how strongly you throw, Hand-thrust Arrows tend to fall near your feet. In those days, many retainers of Daimyo in Western Japan made these short *Uchiya*, Striking Arrows, and carried them on the hip while on horseback. When surrounded by low-level

[18] The *Man'yoshu* 万葉集 *Collection of Ten Thousand Leaves* is the oldest collection of Classical Japanese Waka poetry. The collection was compiled sometime after 759 AD during the Nara period.
The *Taiheiki* 太平記 *Chronicle of Great Peace* is a medieval historical epic written in the late 14th century and covers the period from 1319 to 1367.

Samurai on foot, the mounted Samurai could throw this weapon down from above, thereby creating a helpful opening. The Samurai could then leap down from his horse and begin attacking.[19]

Essential Principles of Military Archery[20] contains the following passage

> …As for *Uchine*, Throwing Arrow, it can be written in two ways, either with the Kanji 討根 attacking + shaft or 打根 striking + shaft. Generally speaking, Throwing Arrows can be just as effective as arrows shot from a bow.
>
> Construction of an Throwing Arrow consists of a blade that is from 5 to 11 inches, according to your desired length. The shaft should be from 1 foot 2 inches to 2 foot 6 inches long, as preferred. The length of the arrowhead should conform to the length of the standard arrows you use.
>
> Make the shaft out of hardwood, just as if you were making a spear. You can also paint the shaft with lacquer. You can use feathers from either large or small birds as you deem appropriate.
>
> In some cases, there are also people who make Uchine without feathers and instead attach a long cord that is tied to the upper arm.
>
> There are various methods of employing these Throwing Arrows.

Also,

> When in combat, you should be single-minded in your focus on stabbing so you penetrate your enemy's armor.

---

[19]*Nihon Bujutsu Meika Den* 日本武術名家伝
Biographies of Notable Masters of Japanese Martial Arts
Iijima Tadaichi 飯島唯一 (?~?)
1902

[20] *Kyuho Gunyo* 弓法軍要
Essential Principles of Military Archery
Miyabe Keijun 宮部継潤 (1528~1599)

Also,

> When large armies have closed and the battle has turned to a melee, spears and swords cannot be employed effectively. Thus when close-quarters fighting begins, nothing surpasses the Throwing Arrow. At the time Samurai spent a great deal of effort developing their Throwing Arrow technique, more so than spear fighting. Also Shuriken were employed. So the story goes.

There are other documents containing information about the Throwing Arrow however these will not be included. Typically an Throwing Arrow will be made of bamboo 1.75 feet long and 0.7 inches thick with a triangular spearhead attached. Four feathers are attached to the base. It can be either thrown or used to stab.

Currently, training in this art is done in a sub-school of the Heki School of Archery. Kyudo Instructor Ouchi is teaching this art today. Two Throwing Arrows from his collection are pictured on the following page.

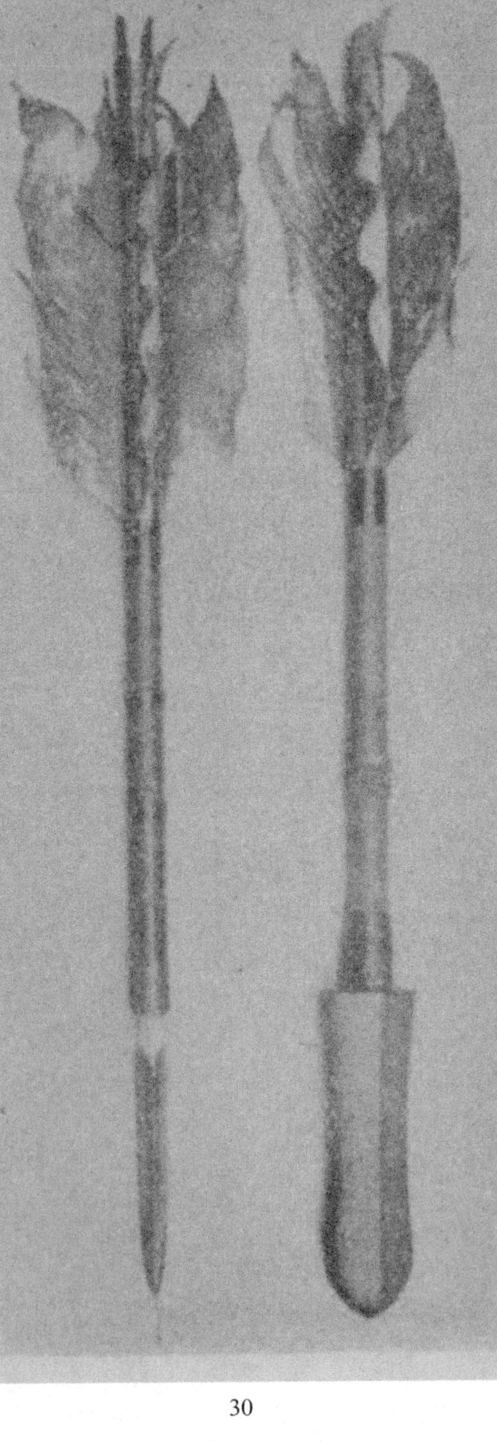

打　根（弓道範士　大内義一氏藏）

（長さ）二尺二寸（莖の徑）六分五厘（穂先の長さ）四寸八分五厘（穂の徑）五分五厘

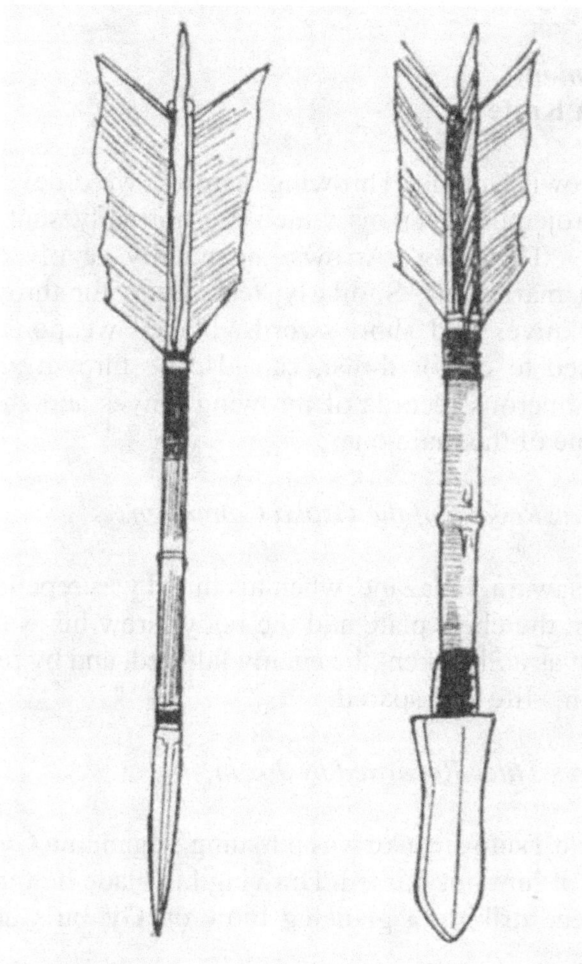

Part of the Uchine	Traditional Units	Feet & Inches
Total Length	2 Shaku, 2 Sun	2 ft 2.25 in
Stem Diameter	6 Bu, 5 Rin	0.78 in
Tip Length	4 Sun, 8 Bu, 5 Rin	5.79 in
Tip Diameter	5 Bu, 5 Rin	0.66 in

Tanto Uchimono
Throwing a Knife

Hand-thrown Uchine, Throwing Arrows, were developed from arrows, a projectile weapons which are normally shot from bow. Over time, Throwing Arrows eventually evolved into an independent martial art. Similarly, techniques for throwing *Tanto Wakizashi*, knives and short swords, bladed weapons which are typically used to cut or thrust, can also be thrown at an enemy Samurai. Numerous records of throwing knives and short swords, here are some of the main ones:

From the *War Record of the Osaka Campaign,*

> Ogasawara Tadazane, when his thrust was repelled by armor between the chest plate and the body, drew his wakizashi and struck it as a Shuriken; the enemy faltered, and by removing his armor, his life was spared.

From *Various Tales Recorded by Jozan*,

> While Tsuto Jinsuke was pursuing Suganuma Gyobu he shot arrows at him but missed. Drawing his blade he threw it like a Shuriken, striking a glancing blow on Gyobu's head. So the story says.

Also from *Various Tales Recorded by Jozan*,

> Yoshihiro fashioned one from black leather. He wore a helmet adorned with bear fur, he fought Inoue on horseback with a sword with a long three-foot blade. Though he was thrown from his horse, he drew his Wakizashi and threw it like a Shuriken, striking Inoue in the thigh on his bow-hand side. In that moment, another Samurai, Oguri, closed in and began grappling with Yoshihiro, eventually taking his head. So the story says.

A book called *A Guide to Japanese Readings of Chinese Words*[21] from around the Meiwa~Anei eras 1764~1781, contains an entry for the word Shuriken:

> The word Shuriken しゅりけん is written as 手裡剣 in Kanji. The Kanji are Hand + Reverse (of the hand, palm) + Sword. It is used in *Gekken no Jutsu*, sword fighting techniques

Later on, the *Dictionary of Written Words and Characters*, there is an entry for Shuriken in the *Seven Tools and Materials* section. [22] This indicates that by the 1700s, the middle of the Tokugawa period, Shuriken techniques had already developed as a variation or sub-category of *Gekken*, sword dueling.

As for the definition of "Shuriken," it is a weapon thrown by hand that combines the functions of both arrows and swords. Its shape and weight have been established through trial and error. However, the specific striking methods, that is, the proper way of throwing, have not been clearly documented.

Uchine, Throwing Arrows, use a *Jikkida*, Straight Throw. However, it seems there were two methods for throwing blades. The first was one was the *Jikkida*, Straight Throw, where the blade is thrown with the tip facing forward. *Various Tales from a Refined Recluse* records,

> A man named Takemura Musashiko, also known as Kouemon, was the son of a man who possessed an unparalleled level of swordsmanship. Takemura was a highly skilled Shuriken practitioner. He could hit the seed in the center of a peach floating down a river with a fifteen-inch blade.

Further, while knives can be throw with the blade forward, they can also be thrown with the handle forward, a reverse of the standard method. This is known as *Kaitenda*, Rotating Throw.

---

[21] *Wakun no Shiori* 倭訓栞 1830
Compiled by Tanikawa Kotosuga 谷川士清 (1709~1776)

[22] *Wakan Onshaku Shogen Jiko* 和漢音釋書言字考
Compiled by Makishima Akitake 槙島昭武(1675~1732)

This latter method is known as *Shiri-ken*, Throwing With the Butt of the Knife. The same word can also be pronounced *Sha-re Ken*. In the Yagyu School book, *Three Questions and Answers About the True Weaving,* [23]

> Other schools have Shuriken that are small blades about three or four inches in length. They are concealed in the palm with the blade facing down and thrown at an opponent. They can also be used without throwing, much like a sword or spear.

These two examples both record the correct way to throw Shuriken. Regarding throwing, obviously schools that teach Uchine, Throwing Arrows, will use *Chokuda*, Straight Throw, but this method is also taught in both Negishi School Shuriken as well as Katori Shinto School Shuriken. However, the Straight Throw of Negishi School Shuriken derived from the needle-shaped Shuriken technique developed by Kadono Sensei.

In conclusion, practical Shuriken throwing techniques developed on the battlefield during the calamitous 1500s and by the early Tokugawa era (1600) had developed into its own martial art. Further, the use of throwing arrows, knives and short, thin spears that somewhat resemble fire tongs, in combat can be found in books like *Diary of Ancient Times in the Warrior's Forest* , which contains the following description,[24]

> For example, both now and in the past, at the outset of a battle, after the first spearmen have made contact, employing Shuriken techniques can result in spectacular victories.

Reading this line it clearly reflects the true situation at the time.

---

[23] *Honori Sanmondo* 本織三問答 *Three Questions and Answers About the True Weaving* by a senior student to Yagyu Muneyori.

[24] *Burin Oseki Nikki* 武林往昔日記 *Diary of Ancient Times in the Warrior's Forest* (referring to Edo.) A collection of anecdotes regarding the customs and events that took place in Edo.
Written by Matsui Shikisui (1661~1733.)

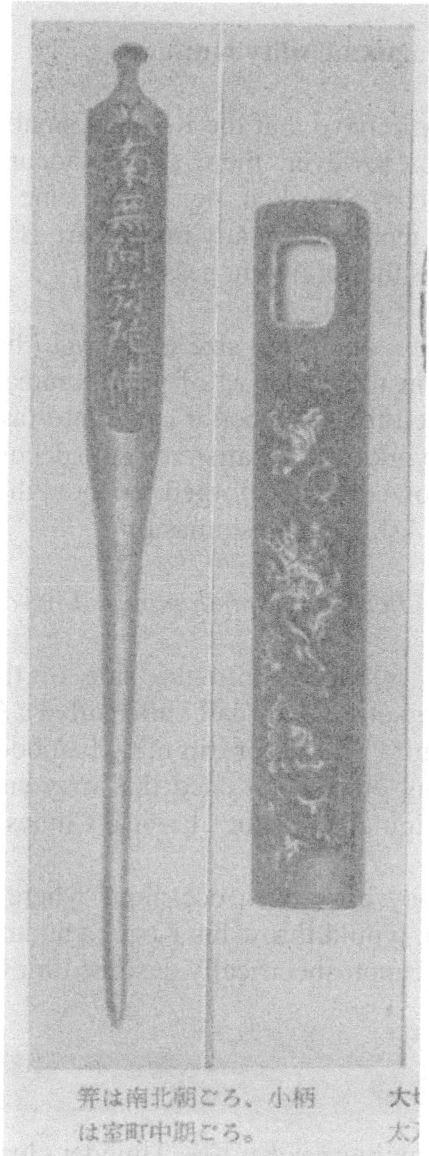

Left: a Kogai hairpin from the Nanboku Era 1336~1392
Right: a Kozuka from the mid-Muromachi era 1350~

Kozuka ni Tsuite
Regarding the Kozuka Utility Knife

It is commonly believe that the Kozuka, small utility knife, is a type of Shuriken. However, there are no documents stating that Kozuka were used as Shuriken, nor has anyone actually observed Kozuka throwing technique in any martial art. *A Brief Explanation of Swords* contains the following passage,

> A small knife, about the size of a *Kogai* hairpin, is inserted into a groove in the scabbard of your Katana.[25] I do not know what this is used for nor when it came into fashion. However if there is metal work by the craftsman Yujo decorating the *Kozuka*, then you know it was forged before the time of Lord Higashiyama (Ashikaga Yoshimasa.)

According to the *Illustrated Sino-Japanese Encyclopedia*,

> A groove is carved into the side of the scabbard of a Samurai's Kanana and a small knife called a *Kozuka* is inserted there. This knife is used for trimming bamboo or wood, cutting paper and any other such need that may suddenly arise. (Of course the original text is in Classical Chinese.) [26]

While there may have been occasions where a Samurai, in an extreme situation, would throw his Kozuka at an enemy, it doesn't appear it was a weapon specifically designed for throwing.

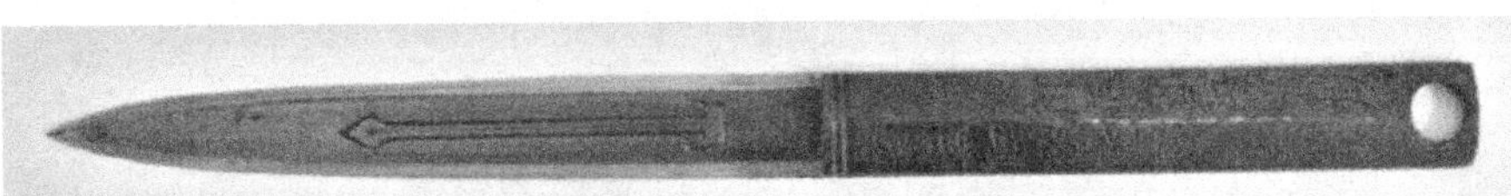

Kaminari-yoke 雷除 Thunder Guard

One type of Kozuka utility knife is the *Kaminari-yoke,* Thunder Guard, which is shaped like a blade with a hole at the base. I tested

[25] Referring to works by Goto Yujo 後藤祐乗 (1440 ~ 1512.)

[26] Brackets by the author.

throwing these like Shuriken, and found that by using a Straight Throw they are effective within nine feet. However, with Rotating Throw the accuracy was inconsistent, with the blade sometimes sticking and sometimes not.

***Tanto Gata* Short Swords Used as Shuriken**
From the collection of Lord Tokugawa Muneyoshi (1897~1989)

The book *War Record of the Osaka Campaign* describes a short sword-style Shuriken. This weapon seems to be the one developed and refined by the lord of Kokura Castle Ogasawara Ukon-daifu Tadamasa (1596~1667) (later known as Tadazane.)

Similarly, the *Record of the Niten School* records an episode where Miyamoto Musashi (1584~1645) fought Shishido Baiken, a master of the *Kusari Gama*, Chain and Sickle. It is said that Musashi defeated him by throwing his short sword as a Shuriken, breaking through Shishido's defenses.

Further, Miyamoto Musashi resided at Ogasawara Castle for several years beginning from around the 11th year of Kanei (1634.) It is likely that since Musashi and Lord Ogasawara shared a profound mutual respect, they would also have discussed martial arts techniques.

Also taking into account the episode in the *Miscellaneous Tales of Koan*,[27] where Musashi's adopted son, Koemon, threw a Tanto, short sword, like a Shuriken at a peach floating in the river. His blade struck true and the point of his Tanto imbedded itself in the heart of the seed at the center.

This serves as a prime historical example of *Tanto-Uchimono*, throwing a short sword, or, in other words, a *Tanto-gata* Shuriken, a short-sword shaped Shuriken. Although the how the blade was thrown remains unclear, it was almost certainly a *Choku-uchi*, Straight Throw, as inferred from the accounts *War Record of the Osaka Campaign, Record of the Niten School* and *Miscellaneous Tales of Koan.* That being said, the striking distance was likely within a range of about two Ken, or twelve feet.

---

[27] *Koan Zatsuwa* 幸庵雑話 *Miscellaneous Tales of Koan*
Watanabe Koan 渡辺幸庵（1582?~1711?)

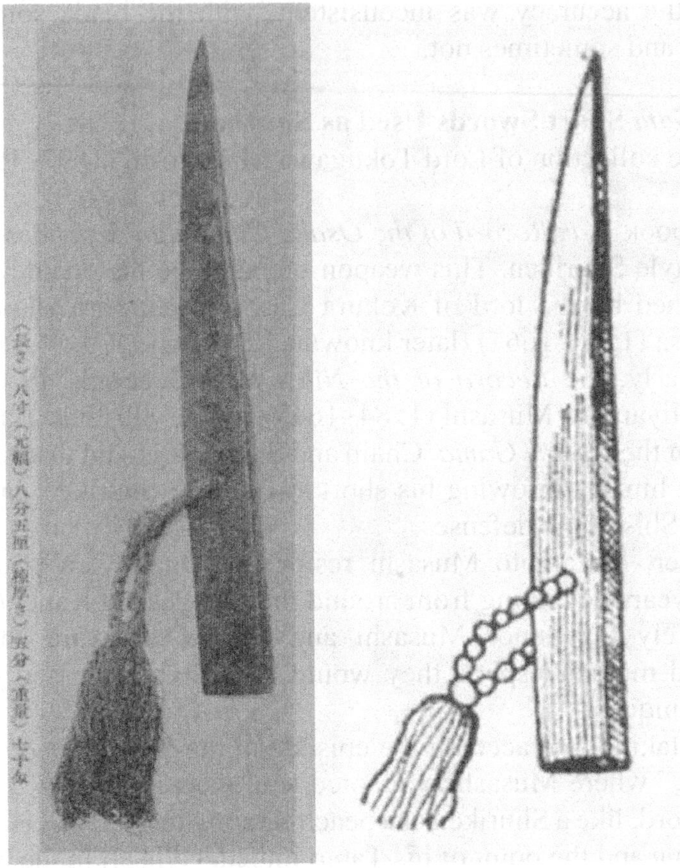

Tanto-gata Shuriken

Category	Japanese Unit	Inches/pounds
Length	8 *Sun*	~9.54 in
Base Width	8 *Bu* 5 *Rin*	~1.01 in
Spine Thickness	5 *Bu*	~0.60 in
Weight	70 M*onme*	0.58 lbs

Kugi Gata
Nail Shaped Shuriken

The Nail Shaped Shuriken are said to have been favored by the last Shogun, Tokugawa Yoshinobu (1837~1913.) According to accounts from the Furusawa family, who were royalty, neither the name of the school of Shuriken the Shogun practiced nor the name of his teacher are known. However Yoshinobu was extremely skilled in this art. Every morning and evening, without fail, the Shogun would tie on a Hakama above his everyday clothing and train until sweat was pouring down his forehead.

He employed a *Chokuda,* Straight Throw, at a distance of about three Ken, eighteen feet, and his accuracy was remarkably precise.

Later, even while living in retirement in Shizuoka Prefecture, he did not abandon this practice, but continued training constantly, even going so far as to order friends from the old days to train until they were proficient. It is said that, during practice, the role of retrieving the Shuriken was performed by Mr. Furusawa.

In addition to the three examples shown at upper left in the photograph on the following page, there are two more with the same shape. The measure from 5 Sun 1 Bu to 5 Sun 6 Bu, with a thickness at the base from 2 Bu 5 Rin to 4 Bu and weighing from 15 Monme to 32 Monme.

It is not clear who used these Shuriken or where they are from, however as they are in the collection of the household of a count, one that was likely linked to the three royal houses with direct lineage to Shogun Yoshinobu, they are no doubt significant. As these Shuriken closely resemble those Yoshinobu favored, it seems likely that there is some connection.

In particular, the inscriptions of smiths are from the *Shinto* period, indicating swords and weapons made in the peaceful Edo era when beauty and craftsmanship became as important as function. This is in contrast to Koto "old period" swords, which were practical and intended for battle.

As the weapons are engraved with the names Nagamichi of Aizu Domain, Owari Fujiwara Sadazane and Nobutaka, we can infer that some of these Shuriken were originally made for members of the Aizu Matsudaira family and the Owari Tokugawa family.

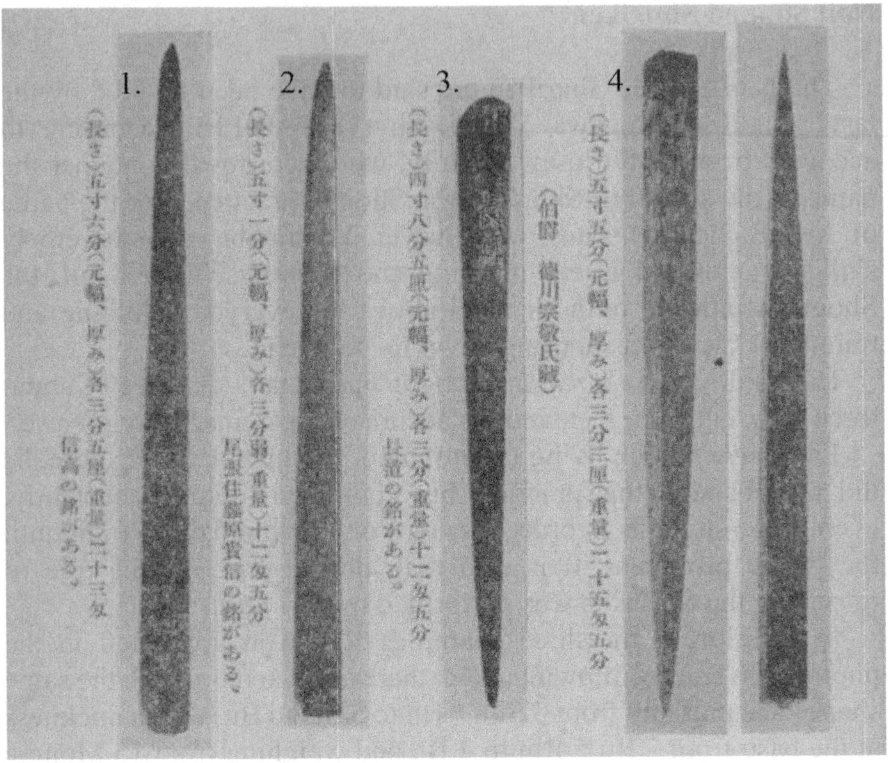

1.
Length: 6.7 inches, thickness 0.41 inches, weight 0.19 pounds
Inscribed with the name *Nobutaka.*

2.
Length: 6.1 inches. Width at base 0.35 inches. Weight 0.10 pounds
Inscribed with Owari-ju Fujiwara Sadazane.

3.
Length: 5.79 inches. Width at base 0.35 inches. Weight 0.10 pounds
Inscribed with Nagamichi.
(Collection of Count Tokugawa Munetaka)

4.
Length: 6.55 inches. Width at base 0.38 inches. Weight 0.21 pounds

Hari-gata
Needle-Shaped Shuriken (Collection of the author) Part I

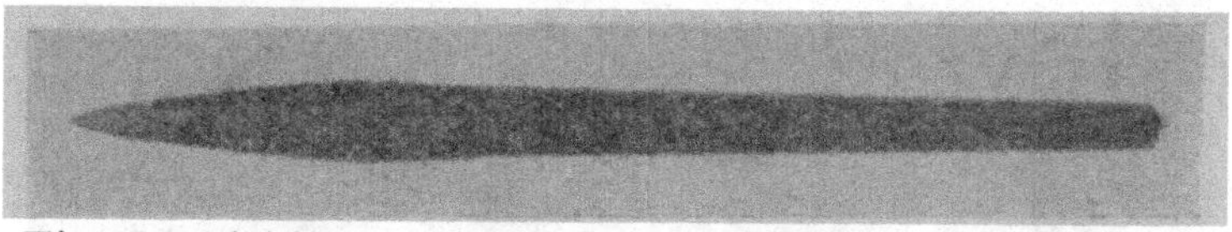

First Negishi Ryu Shuriken developed by Negi Shorei
Length 0.57 inches Weight 1.98 ounces

There is no information regarding the name of the school that uses this style of Shuriken, however they were ones used by Kadono Hirohide, a Samurai from Sendai Domain and date to the Meiwa~Anei eras (1764~1781.) Kadono developed a technique for striking an opponent's eyes at close range.

A detailed record of this can be found in an old book called *Tales of Oshu.* [28] The book is considered to be the pre-eminent true record of Shuriken techniques and was even cited in the thousand volume *Koji Ruien* Japanese encyclopedia commissioned by the Meiji government.

The reason Kadono decided to concentrate on developing a method for throwing *Hari-gata*, Needle-shaped Shuriken, was to find a way to defeat a much larger and stronger opponent. If you blind an opponent in both eyes, the battle is over. No matter how much armor a Samurai may be wearing, no matter how strong his helmet, he must leave openings for his eyes. If you were to advance on an opponent skilled in this technique, you would get done in. It was this concept that gave birth to Kadono's martial art.

This concept the pinnacle of martial arts philosophy, and the technique would later be called *Fusatsu-Sakuto,* ending a fight with a non-lethal strike, and form the basis of Shuriken training. In combat, employing a sword is *Satto Ken,* Killing Sword, however in contrast to this employing Shuriken is *Sakuto Ken* Bring to an Abrupt End Sword.[29]

---

[28] *Oshu Banashi* 奥州波奈志 *Tales of Oshu*

[29] *Fusatsu-Sakuto* 不殺削闘 Ending a fight with a non-lethal strike
Satto Ken 殺到剣 Killing Sword
Sakuto Ken 削闘剣 Bring to an Abrupt End Sword

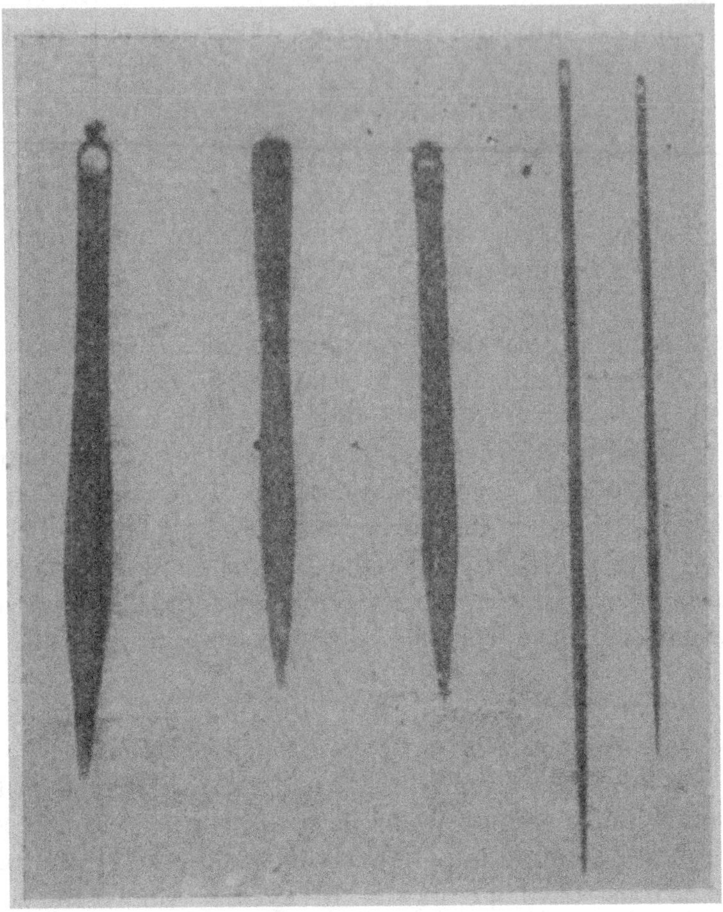

Item	Length	Thickness	Weight
Two Needle-style on right	Short ~5.37” Long ~6.44”	~0.09” ~0.12”	Short ~0.13 oz Long ~0.22 oz
Middle two	~4.41”	Base: ~0.21” Tip: ~0.27”	~0.69 oz
Left Side	~5.25”	Base: ~0.26” Tip: ~0.36”	~1.26 oz

Kadono was a well-to-do Samurai who received an annual stipend of eight hundred Koku of rice, however he would walk about by himself carrying just his Katana. This is because, having devised a method for throwing needle-shaped Shuriken, he would also be armed with eight needles, four concealed on each side of the hair above his ears.

His exquisite needle-throwing technique was at one point lost, however it began to be taught again in Mito Domain and was revived through the ingenuity of Kaiho Hanpei, the second successor of Hokushin Itto School sword.[30] Furthermore, his pupil Negishi Shorei, a Samurai of the Annaka Domain, redesigned it, making the Shuriken longer and thicker.

Later, Tonegawa Magoroku, a Samurai of the Tatebayashi Domain, succeeded Shorei as the head of this school. Thus the school has survived to the present day and its teachings are being transmitted as Negishi School Shuriken.[31]

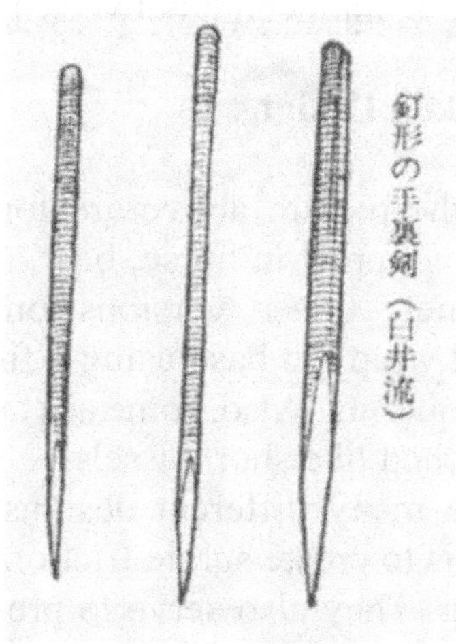

Hari-gata Shuriken
Shirai School

---

[30] Kaiho Hanpei 海保帆平 (1822~1863)

[31] Tonegawa Magoroku 利根川孫六 (1850?~1939)

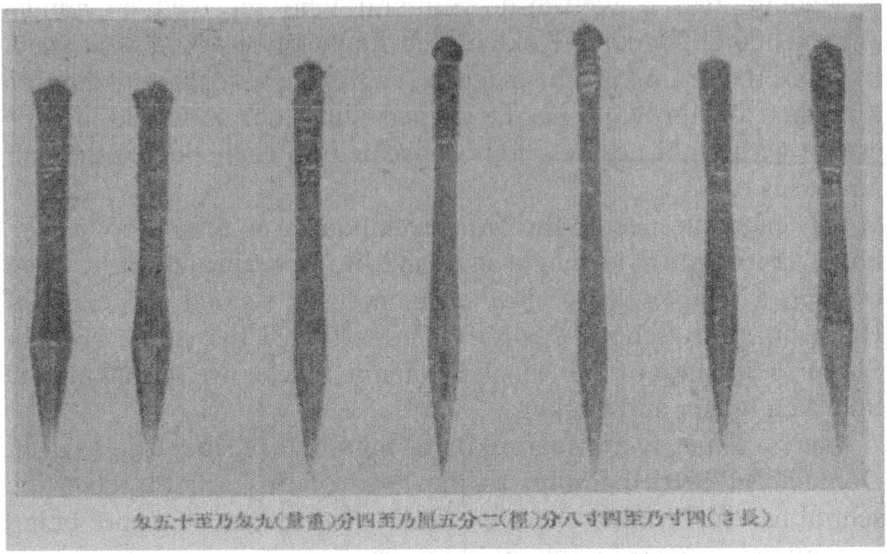

The Shuriken measure 4.8 ~ 5.7 inches in length, 0.3 ~ 0.5 inches in thickness, and weigh between 1.2 ~ 2.0 ounces.

Later Negishi Shuriken Designs

The Shuriken in the picture above are forged in an octagonal shape, with the base wrapped in horse, bear, or wild boar hair and hardened with lacquer. Other versions omit the animal hair wrapping and instead wrap the base using a fine thread for a better tactile grip before lacquering. Also, some are fashioned like needles, with silk threads attached like short tassels.

In short, there are many different designs, the purpose of the wrapping at the base is to create subtle friction points for your hand to use when throwing. They also serve to prevent noise when you shift a Shuriken from your left hand to your right, as well as allowing you to distinguish the tip from the tail by touch alone in the dark.

Furthermore, these wrappings serve the same function as the hilt of a sword.

These Shuriken can typically be thrown a distance of 5~6 Ken, 30 ~ 36 feet. When you throw one at a target, the impact is so powerful that the Shuriken pierces deep as if driven in by an iron hammer, making it extremely difficult to extract. While a master can reach up to 7 Ken, 42 feet, exceeding that range is considered formidable. Throwing a distance of 7.5 Ken, 45 feet, would be equivalent to half the standard range a bow could fire an arrow.

Understand that the two fundamental points in Shuriken throwing are *Kyori*, Distance, and *Kyoryoku*, Power. This portion of the art is relatively easy for beginners to learn. That being said, overall Shuriken Jutsu is a complex art, which may be one reason why the art declined.

As with learning any technique, but particularly with Shuriken Jutsu, you must start at the beginning and gradually progress through all the stages in order. Failing to follow this progression will mean you won't be able to advance your skill at all.

Hari-gata
Needle-shaped Shuriken
Part II

The Shuriken on the right side of the picture on the following page is from the Shirai School. It can be thrown with *Chokuda*, Straight Throw, up to 2 Ken, 12 feet, while distances beyond that, up to about 4 Ken, 24 feet, require *Kaitenda*, Rotating Throw.

You throw by first holding the Shuriken at shoulder height. Step forward with your left foot and releasing.

The Shuriken in the center is from the Katori Shinto School which utilizes a *Chokuda*, Straight Throw, for a distances of up to 2 Ken, 12 feet. When throwing, you stand with your feet together and bring the Shuriken up to *Miken*, the spot between your eyebrows, before throwing.

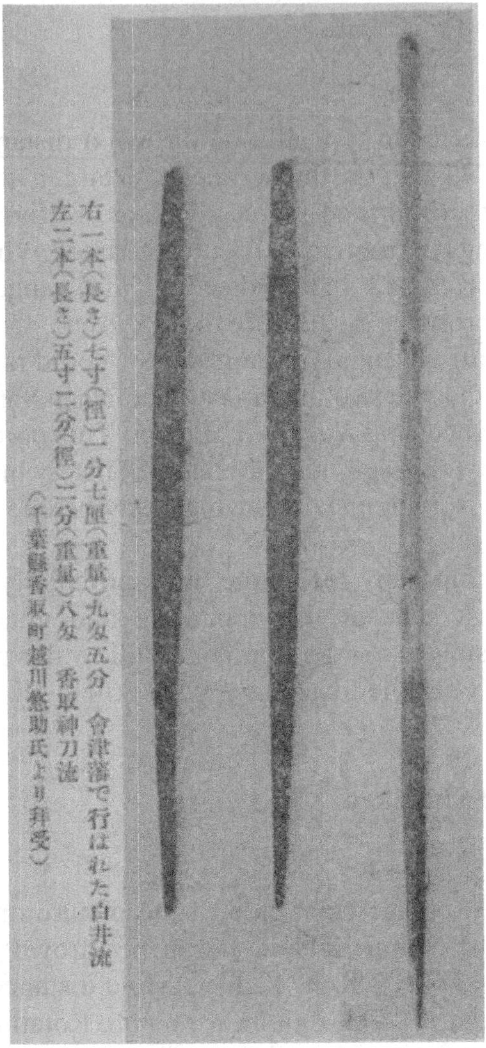

Right: Shirai School of Aizu Domain
Length: 8.6 inches
Weight: 1.3 ounces

Center and Left: Katori Shinto School
Length: 6.3 inches
Weight: 1.1 ounces
Note: Donated by Mr. Koshikawa Yusuke of Katori Town, Chiba Prefecture

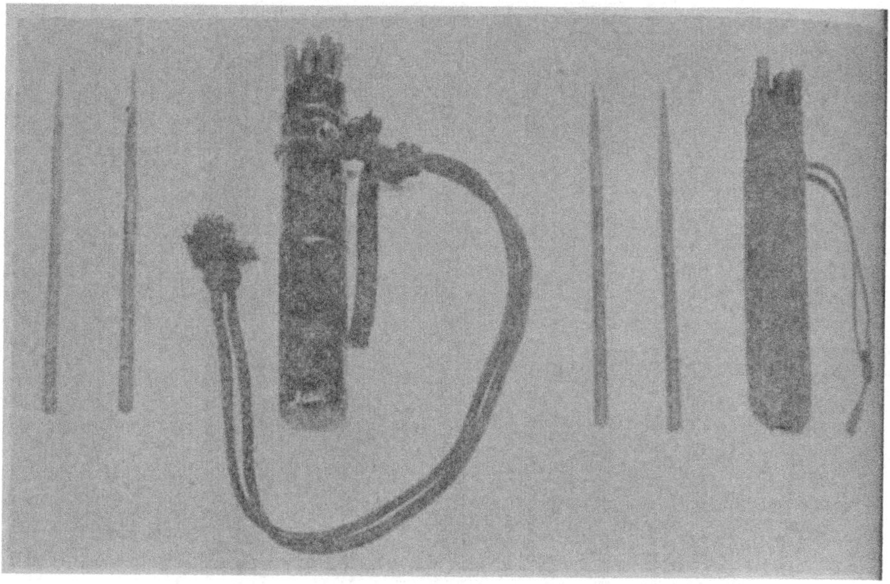

Right: Diameter 0.36 inches. Length 7.2 inches
Left: Diameter 0.36 inches. Length 7.6 inches
Weight unknown

Based on shape, the examples above are probably from the Shirai School. Although these are among the items on display at the Dai Nippon Butoku Kai armory in Kyoto, the name of the school the design comes from as well as how it is thrown is unknown.

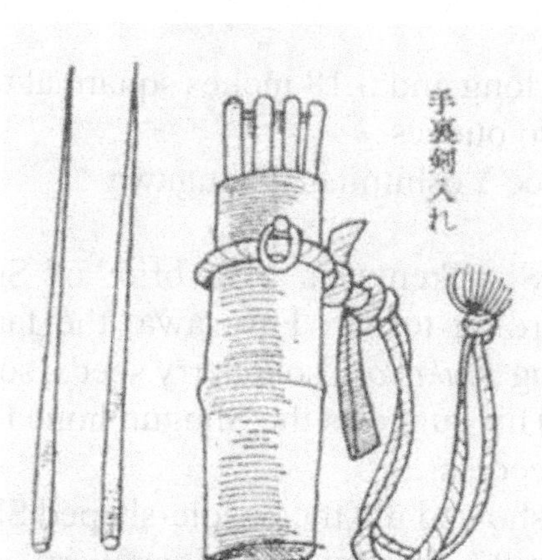

Illustration of a Shuriken Pouch based on the above picture

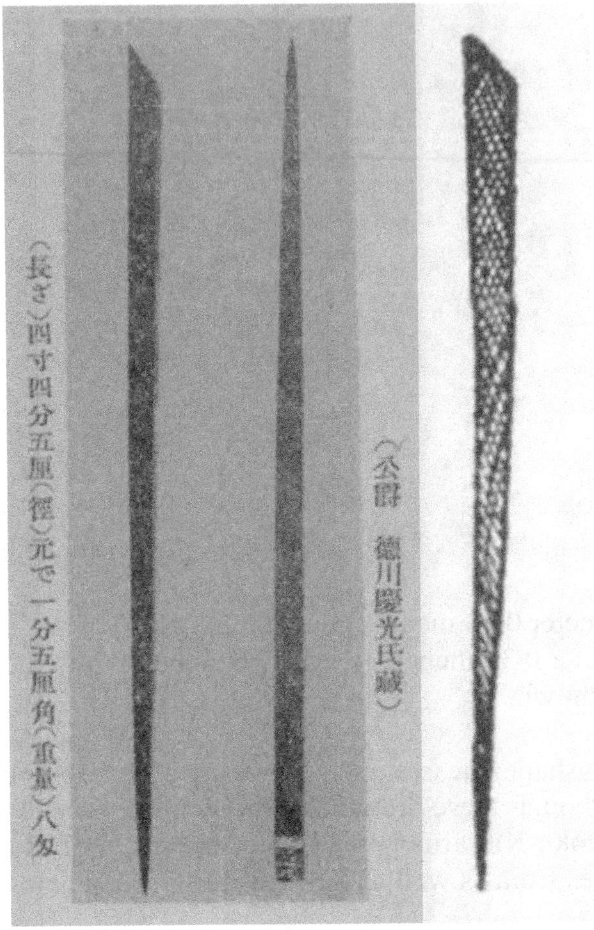

Length 5.3 inches long and 0.18 inches square at the base, weighing approximately 1.06 ounces.
Collection of Prince Yoshimitsu Tokugawa

This style of Shuriken was a favorite of Shogun Tokugawa Yoshinobu. According to Mr. Furusawa, the family steward, the Shogun would hang *Mukuroji*, soapberry seeds, some distance away to use as targets. This suggests the Shogun honed his exquisite skill using these thin needles.

Mr. Furusawa showed me the needle-shaped Shuriken container, and noted that it still held a bundle of soapberry seeds, wrapped up with string.

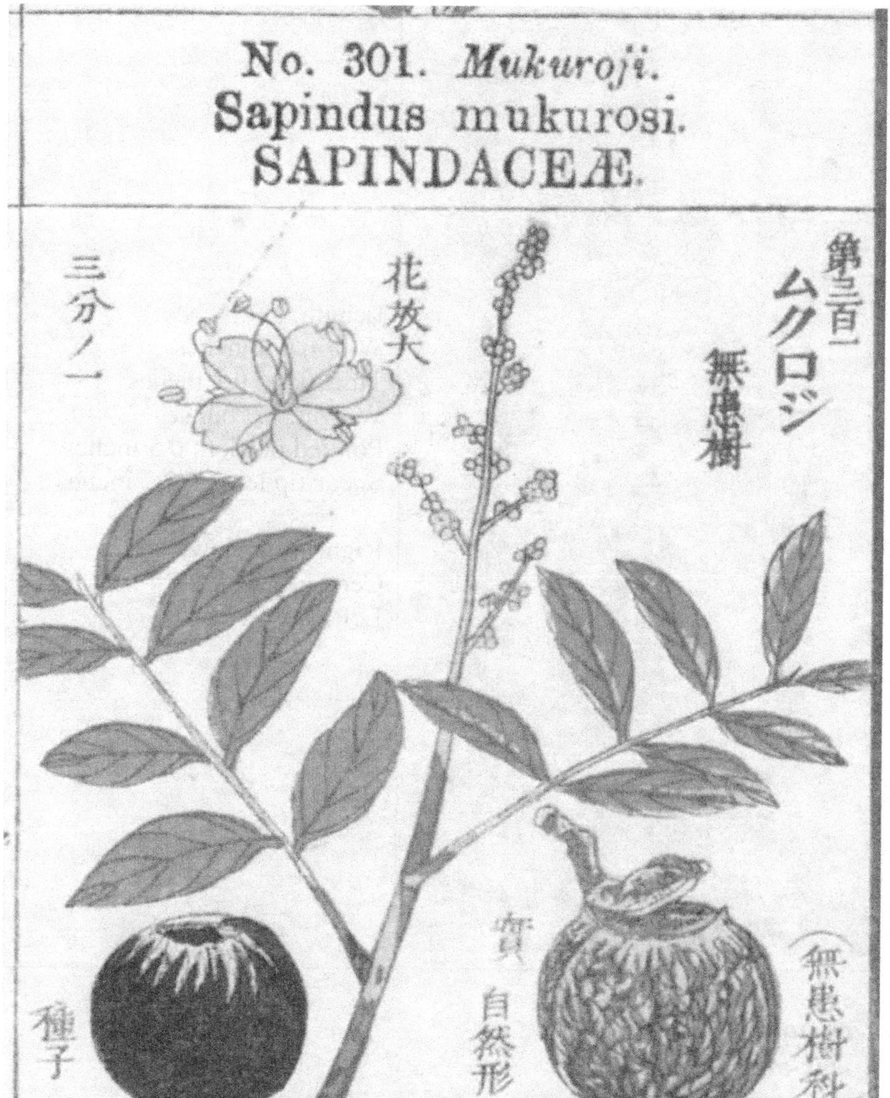

Left: Soapberry seed Right: Soapberry fruit Kanji 無患樹
Yuyo Shokubutsu Zusetsu 有用植物圖説
Illustrated Explanations of Useful Plants
Tanaka Yoshio 田中芳男 (1838~1916)
Ono Motoyoshi 小野職愨 (1838~1890)

Length: 5.7 inches
Width: 0.4 inches
Thickness: 0.2 inches
Weight: 1.7 ounces
Pointed portion: 0.5 inches
Spear-tip length: 2.2 inches

Right: side view
Center: back view
Left: front view

Yari no Ho-gata
Spear-shaped Shuriken
Collection of Duke Tokugawa Yoshimitsu

If you measure back one inch from the point and take a cross section, the cross section forms an isosceles triangle with a 0.4 inch base and 0.2 inch sides. Thus the cutting blade at the end is shaped like the tip of a spear.

While the exact technique for its use is lost, it was a favorite weapon of the last Shogun, Tokugawa Yoshinobu. Its shape suggests it was designed for *Chokuda*, Straight Throw.

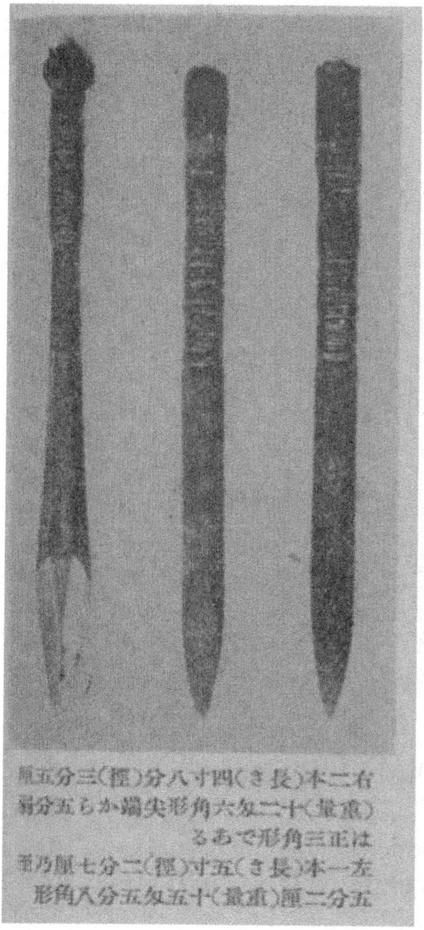

Right and Center:
Length: 5.7 inches
Diameter: 0.4 inches
Weight: 1.6 ounces
Shape: hexagonal

The section extending just under 0.60 inches from the tip is a regular triangular prism.

Left
Length: 6 inches
Diameter: 0.32 ~ 0.62 inches
Weight: 2.1 ounces
Shape: octagonal

Both of these examples are alternate versions of Negishi School Shuriken. The triangular shaped one is from a Negishi School in Mito Domain. The *Hakkaku Gata*, Octagon Shaped, and *Hishi Gata*, Diamond Shaped/Four Sided, Shuriken that are transmitted by the Honden, Main Line, Negishi School.

The *Daho*, Throwing Method, and the effective distance are exactly the same as previously described.

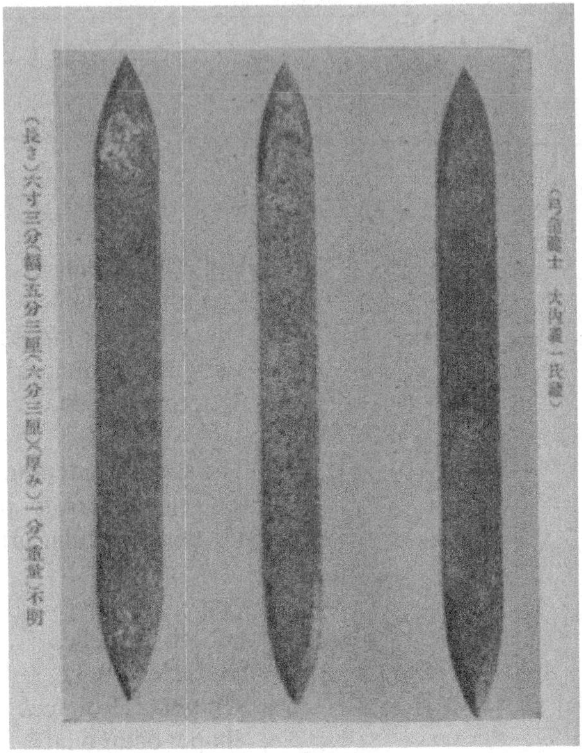

Collection of Kyudo Instructor Ouchi Yoshikazu[32]
Length: 7.5 inches
Width: 0.6 inches (Variant 0.8 inches)
Thickness: 0.12 inches Weight: Unknown

This is *Henpon*, a Flat-type, Shuriken with blades at both ends.[33] This weapon uses *Chokuda*, Straight Throw and its key feature is the fact that you can draw and throw it quickly without having to worry about which end is the front or back. This style of Shuriken was used in Okayama Domain and taught by the Tsugawa School. The Tsugawa School Shuriken are effective up to about 3 Ken, 18 feet.

---

[32] Japanese archery researcher Ouchi Yoshikazu 大内義一 (1880~1956)

[33] Henpon 扁本 flat like something used in book binding.

Juju Gata
Cross-shaped Shuriken
Part I
Collection of Duke Tokugawa Yoshimitsu

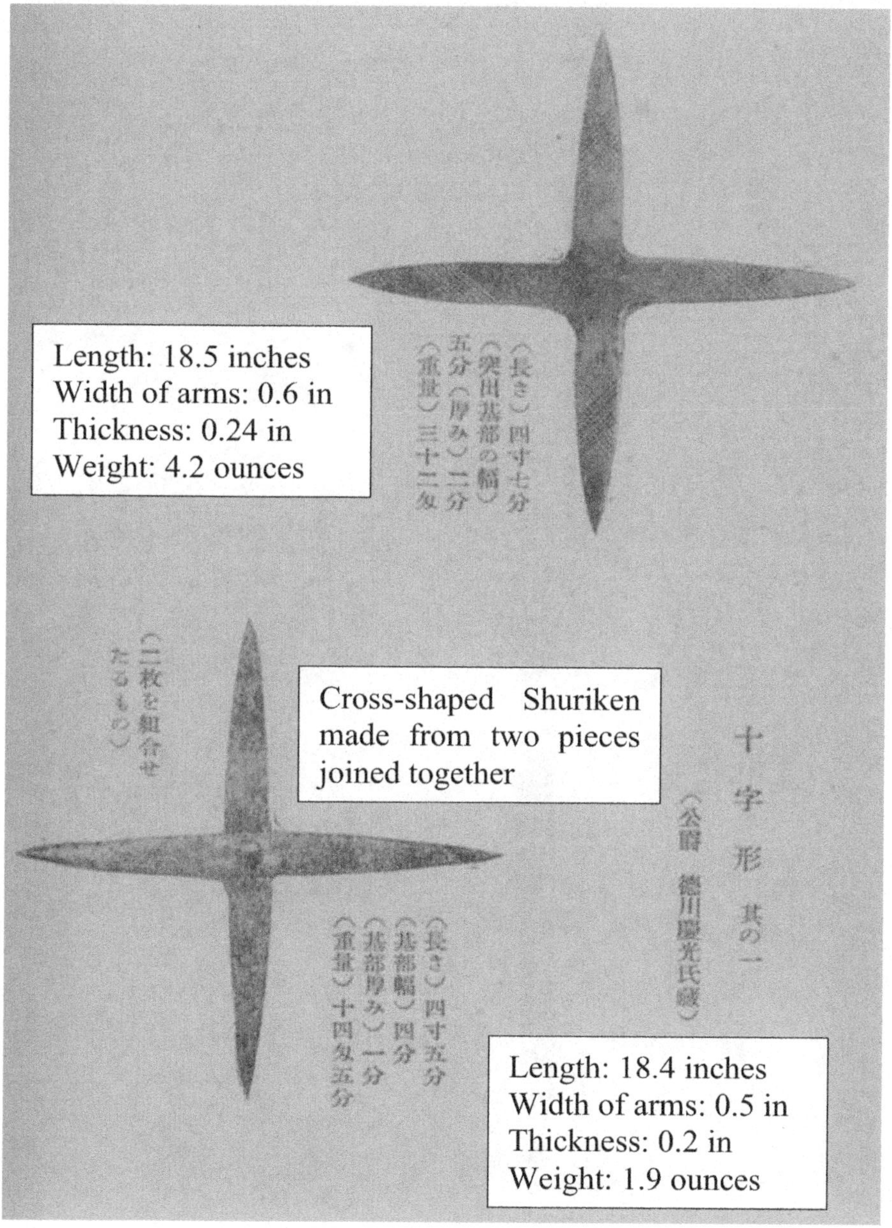

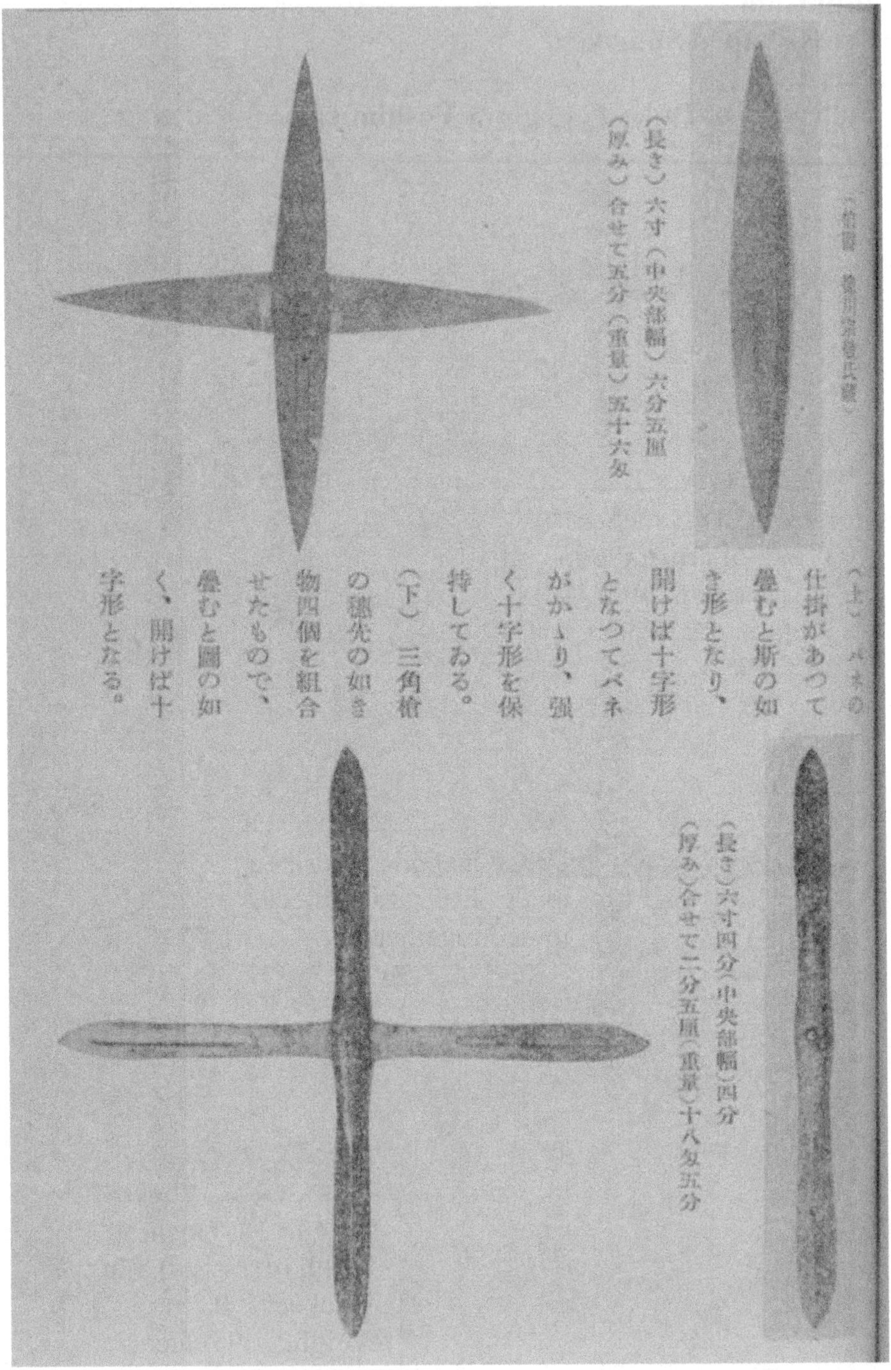

（上）バネの仕掛があつて畳むと斯の如き形となり、開けば十字形となつてバネがかゝり、強く十字形を保持してゐる。（下）三角槍の穂先の如き物四個を組合せたもので、畳むと圖の如く、開けば十字形となる。

（長さ）六寸（中央部幅）六分五厘
（厚み）合せて五分（重量）五十六匁

（長さ）六寸四分（中央部幅）四分
（厚み）合せて二分五厘（重量）十八匁五分

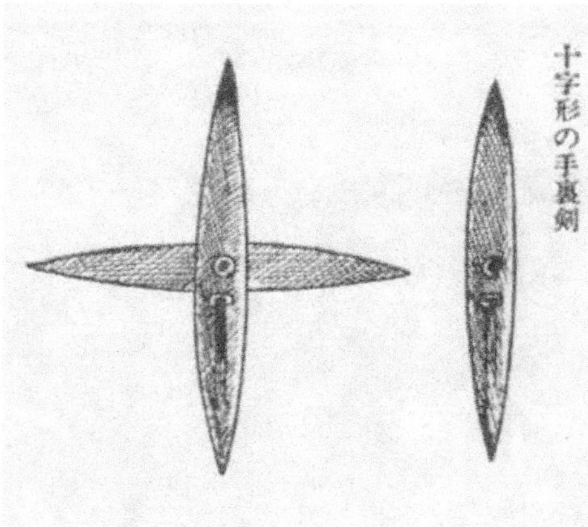

Shuriken from the collection of Count Tokugawa Muneyoshi (1897 ~ 1989.) It is 7 inches long and 0.8 inches across the center. When closed it is 0.6 inches thick and weighs 7.4 ounces.

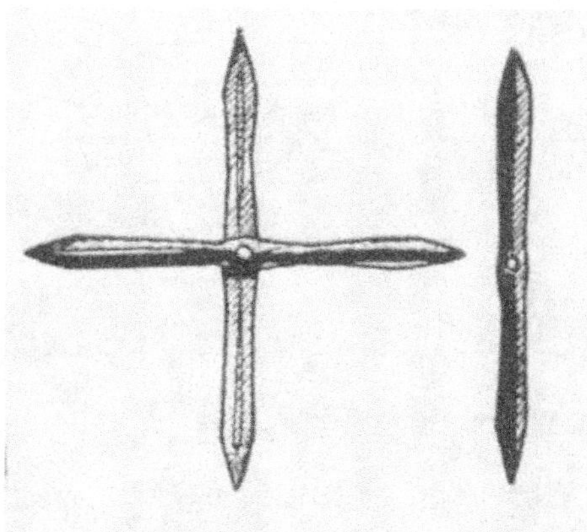

This Shuriken is comprised of four triangular spear-shaped points. When folded it is a single line and when opened forms a cross. It is 7.2 inches long and 0.6 inches across the center. When closed it is 0.3 inches thick and weighs 2.5 ounces.

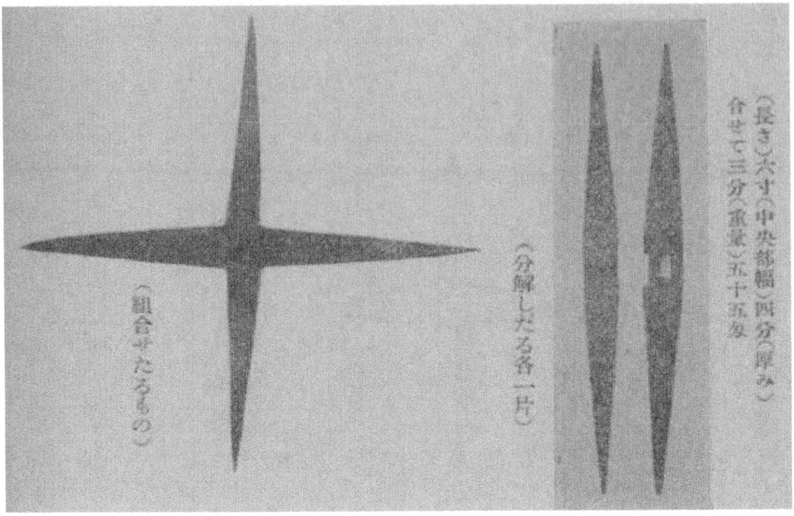

Left: View when assembled. Right: Two pieces separated
Length: 7.1 inches
Width at center: 0.5 inches
Thickness of both pieces overlapped: 0.4 inches
Weight: 7.3 ounces

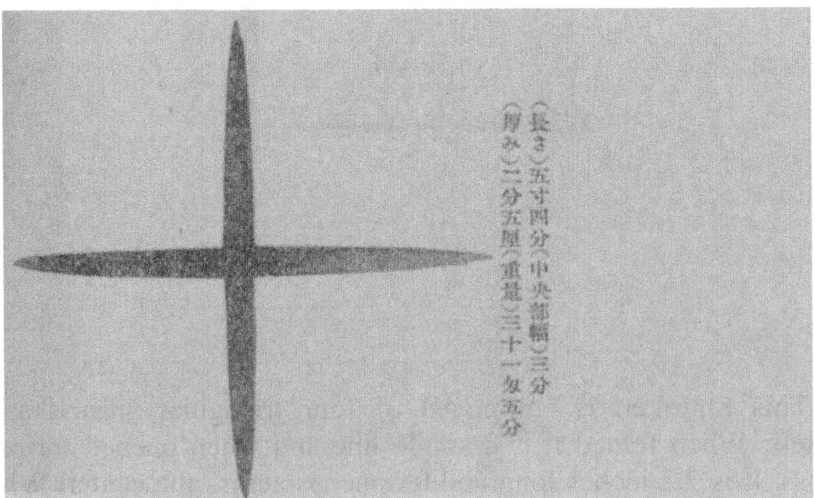

Length: 6.5 inches
Width: 0.5 inches
Thickness: 0.35 inches
Weight: 4.2 ounces.

Cross-shaped Shuriken are thrown at the enemy by making them rotate vertically like a propeller.[34] This technique is the easiest in Shuriken Jutsu, as it can be used freely for both short and long distance. However, because its flight speed is slower compared to other shapes of Shuriken, it has the disadvantage of being easily dodged or struck down.

There is a theory that the cross-shaped Shuriken mentioned in the secret scrolls of the Yagyu School refers to this type. However, another theory suggests the Yagyu documents refers to a Bo Shuriken, Stick Shuriken, with the base carved into a cross shape, while yet another theory claims it refers to both.

There is also a theory that the "cross-shaped Shuriken" that appears in the Yagyu School refers to a certain sword technique rather than the art of Shuriken-throwing itself, but that claim is questionable. The reason is because the following line appears in the *Secret Scroll of the Yagyu School*,

> The secret of the Yagyu School Juji, cross-shaped, Shuriken must never be disclosed to others. In particular, those destined to become the Shogun should be the ones to employ this art.

Furthermore, among the owners of the previously introduced examples of cross-shaped Shuriken, one was a member of the family from which Shogun were selected and another was from the Hitotsubashi family, which, long ago, also had members who went on to become Shogun.

Thus, it seems the late Shogun Yoshinobu refined his Shuriken skills with three types of Shuriken, including those held in the collections of Tokugawa families that inherited the rank of Duke.

---

[34] The author uses the English word "propeller."

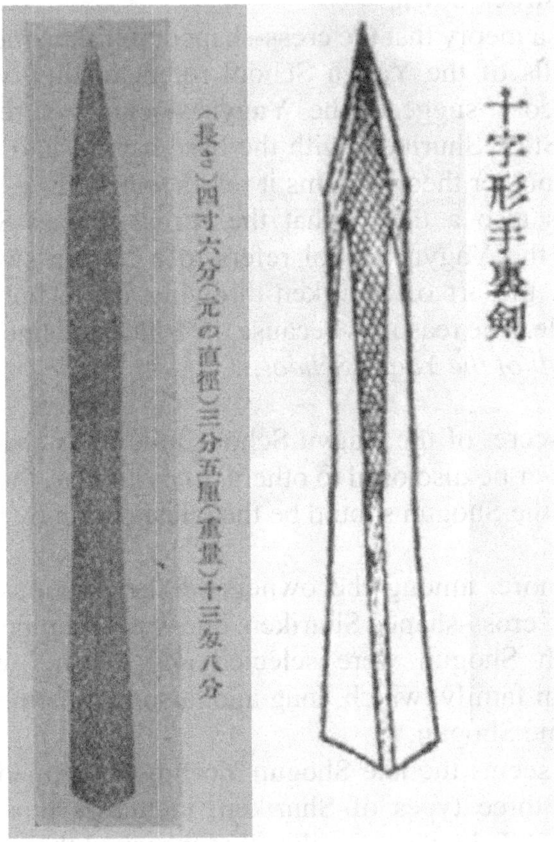

Length: 5.6 inches
Diameter at the base: 0.42 inches
Weight: 1.8 ounces

According to the book *Secret Records of the Yagyu School* there is a weapon referred to as the *Juji Shuriken*, Cross-Shaped Shuriken. In addition the text mentions *Teji Shuriken*, Hand-Shaped Shuriken.[35]

The base of this Shuriken is forged into a cross shape, which is interesting as it aligns perfectly with the aerodynamic principles used in modern aerial bombs and mortar shells. As it moves through the air, the cross-shape of the weapon functions like the fletching of an arrow and stabilize its flight through the air. This effect does not change even though the weapon is made of metal.

Although the *Daho*, Throwing Technique, has not been passed down, experimenting by throwing these Cross-Shaped Shuriken with Negishi School methods reveals they hit with considerable force at an effective distance of somewhat greater or less than 5 Ken, 30 feet. This suggests the weapon may have been thrown in a similar manner.

Members of the Mito branch of the Tokugawa family were famous for mastering this type of Cross-Shaped Shuriken. In the third year of Keio (1867), Lord Tokugawa Akitake went to France as an envoy of the Shogunate (Akitake is from the Shimizu family, one of the three branch houses to the Tokugawa Shogunate, and the current "Flying Tokugawa Shogun" is his great nephew.)[36]

During the trip, Lord Akitake was accompanied by a guard who was a master of *Juji Shuriken*, Cross-Shaped Shuriken. Although the guard carried several Shuriken with him, he did not have to employ them in the peaceful capital of Paris. Before returning to Japan from his travels around England and America, the guard gave the weapons away to various people he had met in France.

---

[35] The Teji Shuriken refers to discerning your opponent's intent, as in "what he has concealed in his hands" By understanding how your opponent will attack, you can counter it.

[36] Brackets by the author. Baron Tokugawa Yoshitoshi 徳川好敏(1884 ~ 1963) Yoshitoshi was a Japanese Army officer known as the "father of aviation in Japan." In 1910, at 27 years old, he became the first person to fly in Japan, piloting a biplane several hundred feet above the Yoyogi Parade Ground in Tokyo

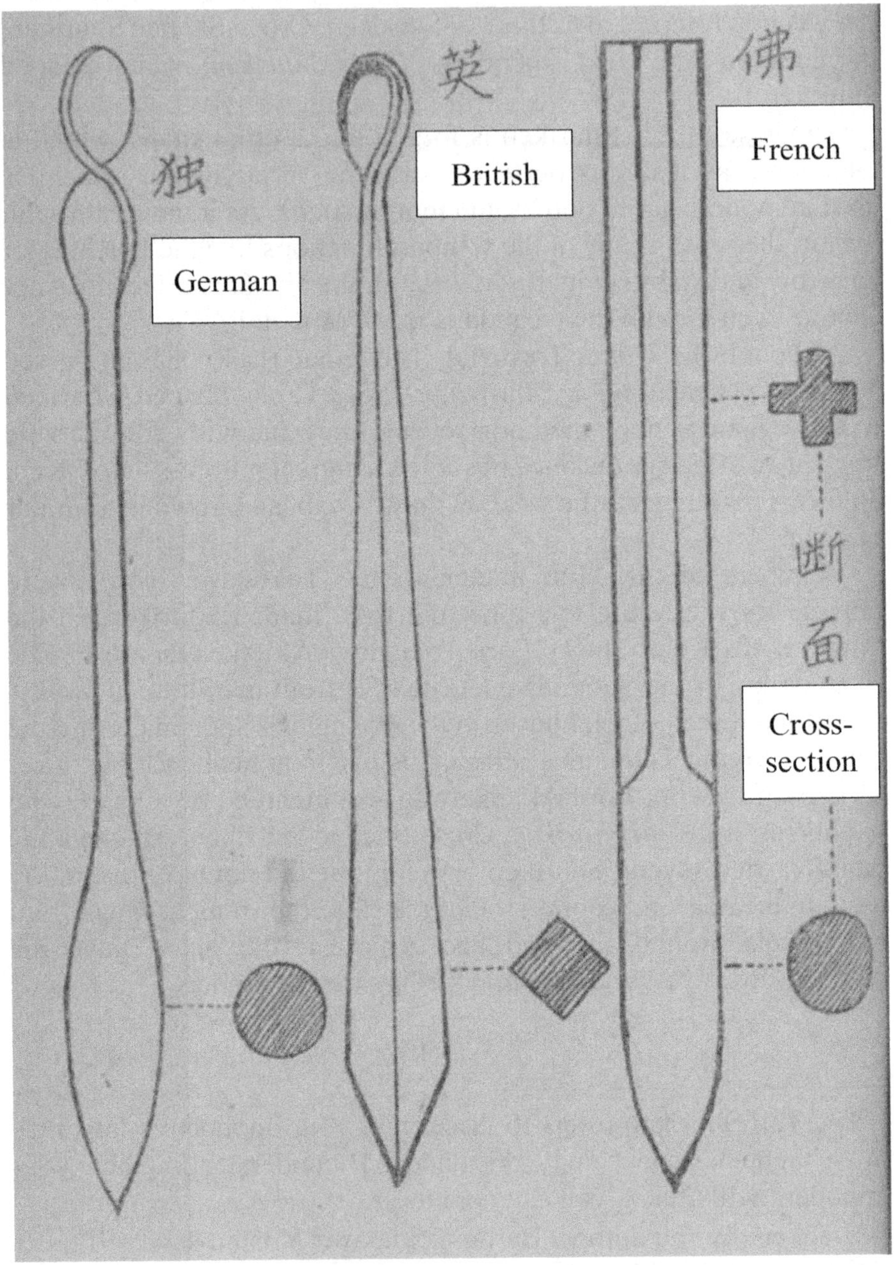

Illustrations of German, British and French fléchettes used during the first world war, from a Japanese encyclopedia.

During the Great European War (WWI), the French army produced a large quantity of metal *Tosen*, or flechettes, and dropped them from airplanes onto German troop concentrations, achieving extraordinary results.

The story goes that an army engineering officer gained the inspiration for this weapon from a cross-shaped Shuriken displayed at a military weaponry museum in Paris. The weapons engineer then copied the shape and conducted field testing. See the illustrations on the previous page.

This cross-shaped weapon was ideally suited for a Christian nation. The ones manufactured by the French army were about 4 inches long, 0.3 inches thick, and weighed 0.53 ounces. A single aircraft could drop between 3,500 and 5,000 of these flechettes. Without needing to fire a shot, they produced the same effect as machine-gun fire. When dropped from an altitude of 1.2 miles, they reached a velocity of 335 miles per hour near the ground. Even a plane flying at the relatively low height of 328 feet, the speed generated was sufficient to be deadly.

Their piercing power was significant. A flechette could enter a mounted soldier's shoulder, pass through his butt, and then penetrate the horse's torso, killing or wounding both man and horse together. The British army later adopted these darts with slight modifications to the shape. In response, the German army developed their own version, increasing the length to five inches and modifying the base to ensure the flechettes stabilized themselves upon release, using them to torment the British and French forces in return. However, as the war was coming to a close, both sides ceased their use after only a short period.

However, even in the present conflict of the Second European War, it is reported that German forces were the first to employ these Shuriken-like fléchettes again. This was reported in a London telegram published in the *Tokyo Nichi Nichi Shimbun, Tokyo Day to Day Newspaper* on May 18, Showa 15 (1940.)

According to the article, the newer versions were 3.6 inches long, making them significantly shorter than the previous version. Unfortunately, nothing is known about whether the shape was refined or changed. However, when the article discusses attacks on British merchant ships, it notes,

This weapon, which was inspired by the Japanese *Shuriken* and was used in the previous Great War, has once again been revived.

We need to abandon the tendency of people consider the only use a thing from the past has is as a resource for historical study. Rather, we should recognize the potential for making new discoveries by examining things from the past. Given the current global situation, I believe we should seriously reflect on this point.

My Experiences With Shuriken

I have some childhood memories related to Shuriken. When I was in elementary school, my father was transferred to Nagano Prefecture, and we lived by the banks of the Chikuma River. This was in 1905 or 1906, some thirty-five or thirty-six years ago

At that time, we would attach strips of Japanese paper like streamers to *Gosun Kugi*, common five-inch nails, and call them "Shuriken." We would amuse ourselves by throwing them straight into wooden targets. There was even a teacher, who was in charge of another class, that would demonstrate Ueda Domain Shuriken techniques. He was said to be descended from Samurai from that domain.

We trained from a distance of about 3~4 Ken, 18~24 feet, from the target. However, if we increased the distance to 5 Ken, 30 feet, even the teacher found it somewhat difficult.

In October, after the rice had been harvested, there was a game called "Netsuki." We would take green bamboo about 0.72 to 0.83 inches in diameter, shave one end to resemble a spear point, smooth out the joints, and pack clay into the tip to give it weight. We made dozens of these bamboo throwing spears, which were a little over a foot long. Then we would practice throwing them into the still-soft soil of the post-harvest rice fields. The challenge was to knock down the bamboo spear of those who had thrown before. Knocking down an opponent's spear meant you got a point. I often won, and remember piling up the collected bamboo spears until they resembled a mountain under the eaves of our house.

It is said that long ago, the Samurai warlord Sanada Yukimura (1567 ~ 1615) taught farmers how to make bamboo throwing spears. He then positioned the farmers on rooftops or other elevated positions and they would hurl the spears them down upon advancing enemy forces like rain, breaking the opponent's charge. This clearly the origin of *Uchine Jutsu*, Japanese Throwing Spear Techniques, as well as *Tanto Uchimono*, Throwing Short Sword or Other Blades.

At that time, there was also a craftsman lodging next door. His name was Hashiba Nishiki, and he too was said to be of Samurai descent from some unknown domain. He was licensed in Hanbo, wooden half-staff techniques, and was also an established Judo practitioner.

I remember how he would throw his 4 Shaku, 4 foot, staff like a spear so it made a sharp "whooshing" as it flew through the air, and strike wild dogs that roamed about causing trouble. His throws were accurate, and the tip of the staff would hit the target straight on.

Translator's Note: Nata

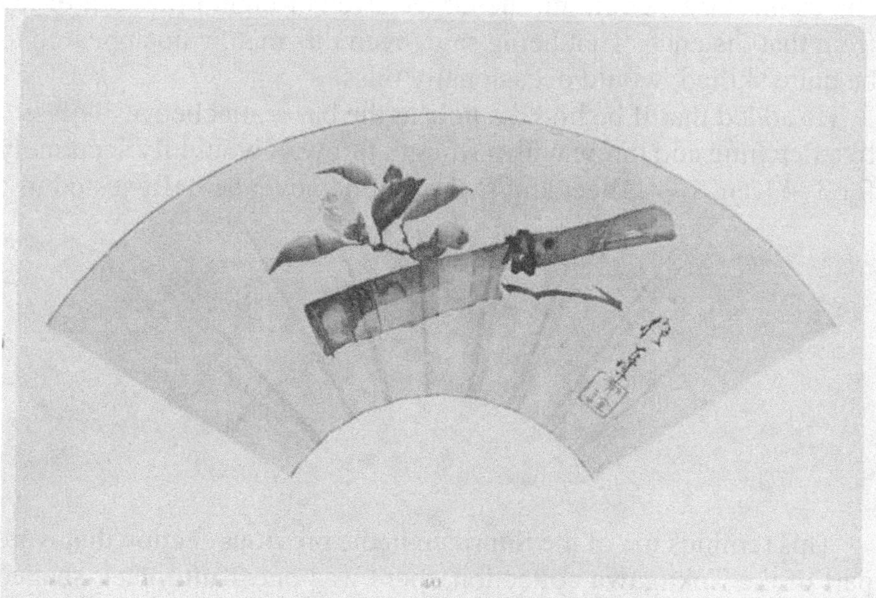

Tsubaki and Nata 椿と鉈図 by Sakai Hoitsu 酒井抱一
Edo Era. The author indicates the man cut the point off Wakizashi swords in order to make the end flat.

This man owned many Wakizashi short swords, and from them he made Nata, multipurpose hatchets, for trimming scraps of wood for carpentry use or kindling. Since the natural tip of six or seven inches of the Wakizashi was unnecessary, he would break it off and set it aside. He then taught me how to use these sword-tips as Shuriken. He once told me,

> When the snow falls, I go into the mountains and live in a mountain hut. I often find that as I am chopping wood with my Nata hatchet, mountain hares come out in search of food because of the snow. I keep working and if they come within range, I throw my Nata at them.

Having told me this story, he demonstrated throwing his Nata from a distance of about 1 Ken, 6 feet. He held the Nata by the blade with the end of the handle facing forward and threw with a Rotating Throw. He said he could strike an object accurately from 6~9 feet and estimated he could hit the target seven or eight times out of ten from that distance. That being said, even this man, who appeared to be quite skilled, would occasionally miss.

He added that if he bored a hole at the base, attached a silk tassel like fletching and threw with a straight throw, it would fly accurately for 3~4 Ken, 18~24 feet and strike, but he never actually tested it.

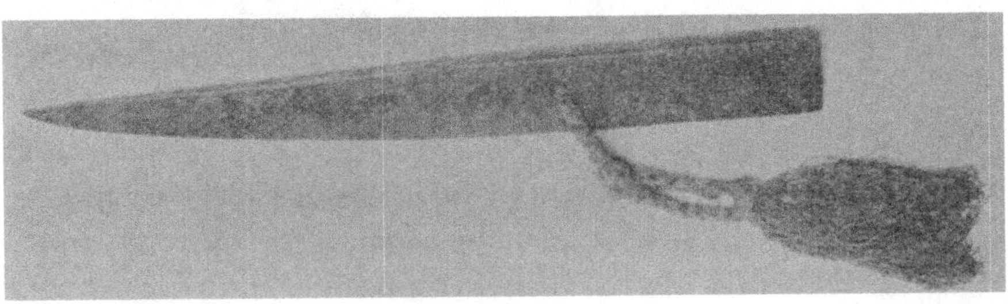

This reminds me of the Shuriken in the previous section that were part of the Tokugawa collection, there was one Tanto-gata, shaped like a short sword, about eight inches long, with a hole in the hilt and a red tassel attached.

With *Chokuda*, a Straight Throw, you can strike a hundred times in a hundred throws, but with a *Kaitenda,* Rotating Throw, perfect accuracy is impossible. This is clear even in theory, and almost all

Shuriken described in written sources are used with a Straight Throw. This is a lesson I learned from my childhood experiences.

Later, on April 15, Showa 2 (1927,) there was a lecture on Shuriken at the Miyama Club in Denma Cho, Yotsuya Ward, Tokyo City. It happened snow heavily that day, remarkable as such snowfall was rare. The lecturer, revered elder Tonegawa Magoroku, the second head of Negishi School Shuriken Jutsu, began introducing Shuriken throwing as its own martial art. At our earnest request, the revered elderly gentleman trimmed thick bamboo chopsticks with a small knife, thereby crafting improvised Shuriken and threw them into Tatami mats. They pierced with tremendous force, and everyone present was astonished.

Two days later, on the 18th, I visited his residence in Sendagaya, formally became his student, and have continued to devote myself to training this for fourteen years up to the present day.

On February 11, Showa 12 (1937), at the Ryogoku National Stadium, a dedication exhibition of Kobudo, classical martial arts, was held as an offering to Kashihara Shrine. Kobudo practitioners from all over the country participated and I also took part, demonstrating Iai Jutsu, Sword Drawing, and Shuriken Jutsu. On that day their majesties Prince Chichibu Yasuhito, the younger brother of Showa Emperor, and Prince Nashimoto Morimasa, the head of the Nashimoto-no-miya house, were in attendance. Revered Master Tonegawa, who was eighty-seven-years-old at the time also came to observe. When I visited him the following day, he critiqued my demonstrations saying, “You have not yet mastered the true art of Shuriken.” His words were, of course of, meant to inspire and encourage me.

The next day, he sent me by post a formally written will, in his own hand, designating me as the successor to the Negishi School of Shuriken. It was signed:

Tonegawa Magoroku Masatoshi
Former retainer of the Tatebayashi in Joshu, Junior Fifth Rank

The document was signed with a Kao, a decorative official signature. When I went to express my thanks, I found that he was confined to bed with a slight cold. He said,

> The Japanese spirit cannot be expressed through logic. It remains embodied in *Kobudo Kata*, Classical Martial Arts Techniques. That is why it is so important that these traditions be revived. If Shuriken Jutsu, which began from something as small as a needle, dies out, it can never be revived. Please, find a way to preserve it for future generations.

Saying this, he taught me one final *Hiden*, secret transmission. I thought it would be a technique, but it turned out to be a method of making Shuriken and how to balance them.

The following year, I was drafted into the army and sent to northern China. Not long after my return, on January sixteenth of Showa 14 (1939,) I received news from his successor, Mr. Tonegawa Hisashi, that my revered elderly master had passed away on December seventh at the age of eighty-nine.

Strangely, enclosed with that notice, General Honjo of the Imperial Army sent me a mimeographed copy of a field newspaper.[37] It was a report regarding Sergeant Maeda Isamu, a grand-student of Tonegawa Magoroku Sensei. The Sergent, while stationed on the central China front, had snuck up on an enemy soldier and killed him using a Shuriken.

On February seventeenth, of the following month, the late master was buried at Shokoji Temple cemetery in Yoshimachi, Maebashi City.

I began studying Negishi-ryu shuriken in Showa 2 (1927) and I also began researching and training other schools of Shuriken, particularly the Shirai School, which is originally from Aizu domain.

When throwing Shuriken in the Shirai School, a Straight Throw is used for distances from 9~18 feet. For distances from 18 ~24 feet a Rotating Throw is used. Beyond that distance, it becomes somewhat difficult to throw with any accuracy.

The *Hari no Jutsu*, needle-throwing technique, of the ancient Kadono tradition, which forms the basis of the Negishi School, is effective to around twelve or thirteen feet, but throwing with the much heavier Shuriken that is the standard Negishi School weapon,

---

[37] Honjo Shigeru 本庄繁 (1876~1945)

you can attack a target as far as seven and a half Ken, almost forty-five feet, which is about half the range of a standard bow. In medium-distance throwing techniques, it stands unquestionably in first place.

The founder of the Negishi School was Kadono Sensei. He was a Samurai of the Sendai domain, and was given the honorific title Guardian of Izu. The man who further refined the tradition was Negishi Shorei, the head sword instructor to the Samurai of the Annaka Domain in Joshu.

Negishi Shorei's nickname as a swordsman was "Little Tengu of Joshu," and his two-handed thrusts were said to be a *Kami Waza*, of almost supernatural skill. In fact, the Daimyo in charge of his domain forbade him from using it outside of combat.

Negishi Shorei's memorial stone
Annaka Town, Gunma Prefecture

In Annaka town, a monument commemorating his virtues has been erected, and his grave is located at Myoko-in Temple in the same town.

The man who further refined the weapons and techniques used in the Negishi School was my aforementioned revered elder master Tonegawa. He was a retainer of Tatebayashi in Joshu, and he fought in the Aizu campaign during the 1868 Boshin War, when he was around eighteen years old. Tonegawa Sensei excelled in the three arts of sword, spear, and horsemanship, and later served for many years as head of one of the counties that made up the domain.

The Shirai School practiced in Aizu domain developed a weapon called *Te-bojutsu,* hand stick techniques. These are sticks of hardwood about one foot and nine inches long, with four iron wires stretched along the length. At the end of the stick is a chain with a six or seven inch blade attached to the end.

Practitioners of this school would engage the enemy using a combination of this *Te-bojutsu* together with Shuriken.

The shuriken themselves were iron rods with tips sharpened needles, about eight inches long and with a diameter of about a quarter inch.

The founder of the school is unknown, referred to only as Shirai, but in the late Edo period a Samurai named Kurokawauchi Dengoro Sakunori, was the head instructor of this school. Kurokawauchi was also skilled in both *Iai*, Sword Drawing, and *Naginata*, Halberd, techniques. He lived in Ichiban-cho, Shinmachi, Aizu, where he also maintained a private dojo. His grave is at Amida-ji Temple in Aizuwakamatsu.

Finally, I have been researching both throwing Shuriken and crafting them for more than ten years. I believe that throughout the country there still remain a considerable number of documents hidden away along with actual examples of Shuriken. Taking this opportunity, I would be overjoyed if anyone could share any information they may have regarding this subject as it would greatly benefit research into Shuriken.

End

Translator's Note: Yagyu School Cross-Shaped Shuriken

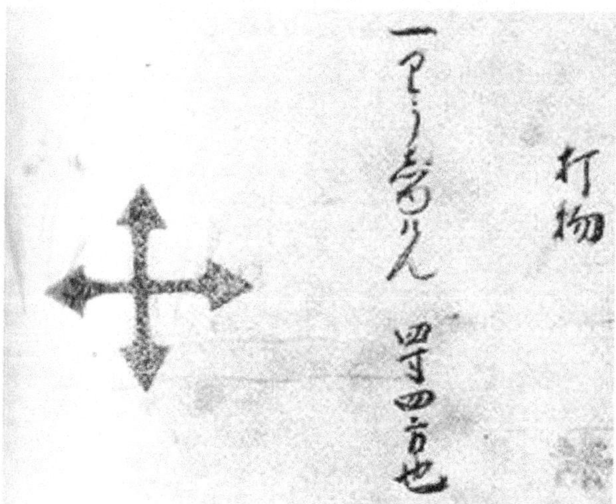

柳生新影流 *Yagyu Shin Kage Ryu*
玉成集 *Lessons on Becoming A Polished Jewel (A Great Human)*
打物 *Uchi Mono* Striking Weapons

This is about throwing a Wakizashi, Shoto or whatever you have on you at the time. You can practice with various weapons turning training. However the weapons often fail to hit their mark, and therefore should not be relied upon.

The Japanese Throwing Arrow

Murakawa Kiyoshi
村河清 (1882~1913)

Excerpt From:
Shoryu Kyudo Shinanka Shokai
諸流弓道指南歌詳解
Detailed Explanation of Instructional Verses of Various Schools of Kyudo
Published 1928

はる〳〵と旅行路や夜の道
要心の時打根はなすな

(日置雪荷)(日置道雪)(大藏)(印西)(竹林)(大和)

雪荷、道雪、大藏、印西、竹林派の歌は路を道要を用させり日置流の解に曰く打根と云ふは篦太くして長さ七八寸より一尺計り羽長さ三四寸根は鏃の笹穂の如きものをすげ筈に腕貫の如き紐をつくこの打根は弓にて射るには非すまさかの時に突く爲のもの也或は矢を打つける爲也矢故に弓射る時は右の紐を腕にかけて射る也委しくは用方の卷に出たりとせり近來の打根を見るに筈に紐を付けす筈巻と末作の間に付せり之は手裏劍の代用又城壁を登るに挿込みて登るに用ひ又敵近づくを突くにも用ふと云へり歌の意は遙々即ち長き旅行の途中夜の道中など用心すべき時には打根を持ちてはなすなとの義也

Uchine, the Japanese Throwing Arrow

When travelling down the road,
Or walking at night,
In times that require vigilance,
Keep your Uchine close at hand

This poem, which can be found in the Sekka Branch, Dosetsu Branch, Ogura Branch, Inzai Branch, Chikurin Branch, Yamato Branch and lineages, emphasizes that while on the *road* it is *required* that you remain *vigilant*.

According to the Heiki School of Archery, the Uchine, Throwing Arrow, has a thick shaft of about 8 ~ 12 inches in long, with feathers roughly 3½ ~ 5 inches long. The head of the Uchine is a spear shaped and resembles a bamboo leaf about 3 or 4 inches long.

There is a simple cord attached to the shaft of the Uchine which is tied to your arm.

When facing an unexpected attack, and you don't have time to nock the Uchine on your bowstring, it can be used to stab. You can also throw it if necessary.

When shooting from a bow, you leave the cord attached to your arm. Details regarding this can be found in the *Yokata*, Application, scroll.

Recently when you see people using Uchine, they often do not have a cord attached just above the nock at the back. The standard method is to bore a hole at the base of the shaft just above the nock, and cord is threaded through.

The Uchine can also be used the same way as a Shuriken and can serve as a tool to help you climb the wall of a castle by stabbing it into the chinks in the stone wall. And, of course, you can stab an enemy closing in on you.

As for the meaning of the above poem, it teaches that while on a journey to a distant domain, especially when walking at night, you should always have your Uchine ready for immediate use.

The Japanese Throwing Arrow

Kitamura Akitaro
北村明太郎 (?~1933)

Excerpt From:
Yumi no Michi ゆみの道
The Way of the Bow
1926

打根の起原 （道雪）

日置彈正是れを工夫すと云ふ。

昔時手突矢と云ひ所謂手鎗の一種とも見るべく、行旅の時陣中常に紐を付け腕首に掛け居り射倒したる時止を刺すに用ひたりと云ふ、又早矢射損じ乙矢を番ふの隙なき場

一一九

The Origin of the Uchine, Japanese Throwing Arrow[38]

It is said this weapon was devised by Heki Danjo.

Long ago there was a weapon known as a *Tezuki-ya*, Hand-thrust Arrow, which may be regarded as a kind of *Teyari,* Hand-spear. When traveling or walking around an army at camp, Samurai carried this with the cord tied just above their elbow.

It was said to be used stab the final blow to an enemy you had shot down with your bow. Furthermore, in cases where your first arrow missed and you don't have time to nock a second arrow before the enemy is within striking distance, you can use the Uchine as a Shuriken to strike him down. It is also called a Shurenken.

---

[38] Heki School 日置流 is an archery school founded in the late 15th century by Heki Danjo Masatsugu 日置弾正正次 His school emphasised a straighforward and practical method of shooting called Busha 武射 that was effective on the battlefield. It emphasizes three core principles known as Hi-Kan-Chu 飛貫中

Hi 飛 To fly
Kan 貫 To pierce
Chu 中 To center/hit

As the method was ideal for Hosha 歩射 archers on foot, it became the primary method used by several branches of the Heki School, including the Inzai Branch 印西派, Chikurin Branch 竹林派, and Sekka Branch 雪荷派 lineages.

一二〇

合敵間近に進み來りたる時手裏劍（手練劍と云ふ）に代用して打止めりと云ふ。

打根の圖（寫眞參照）

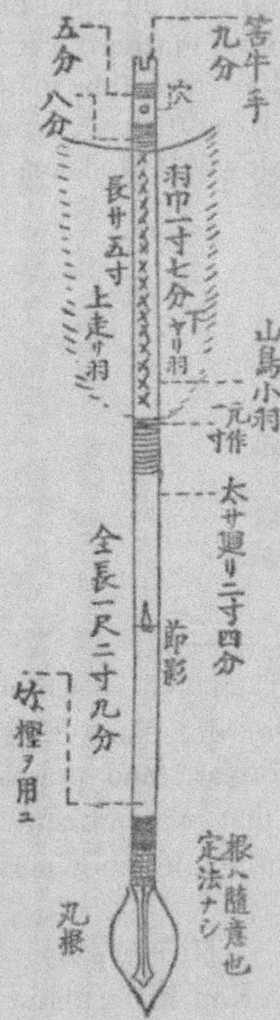

四つ立、四枚羽正面山鳥

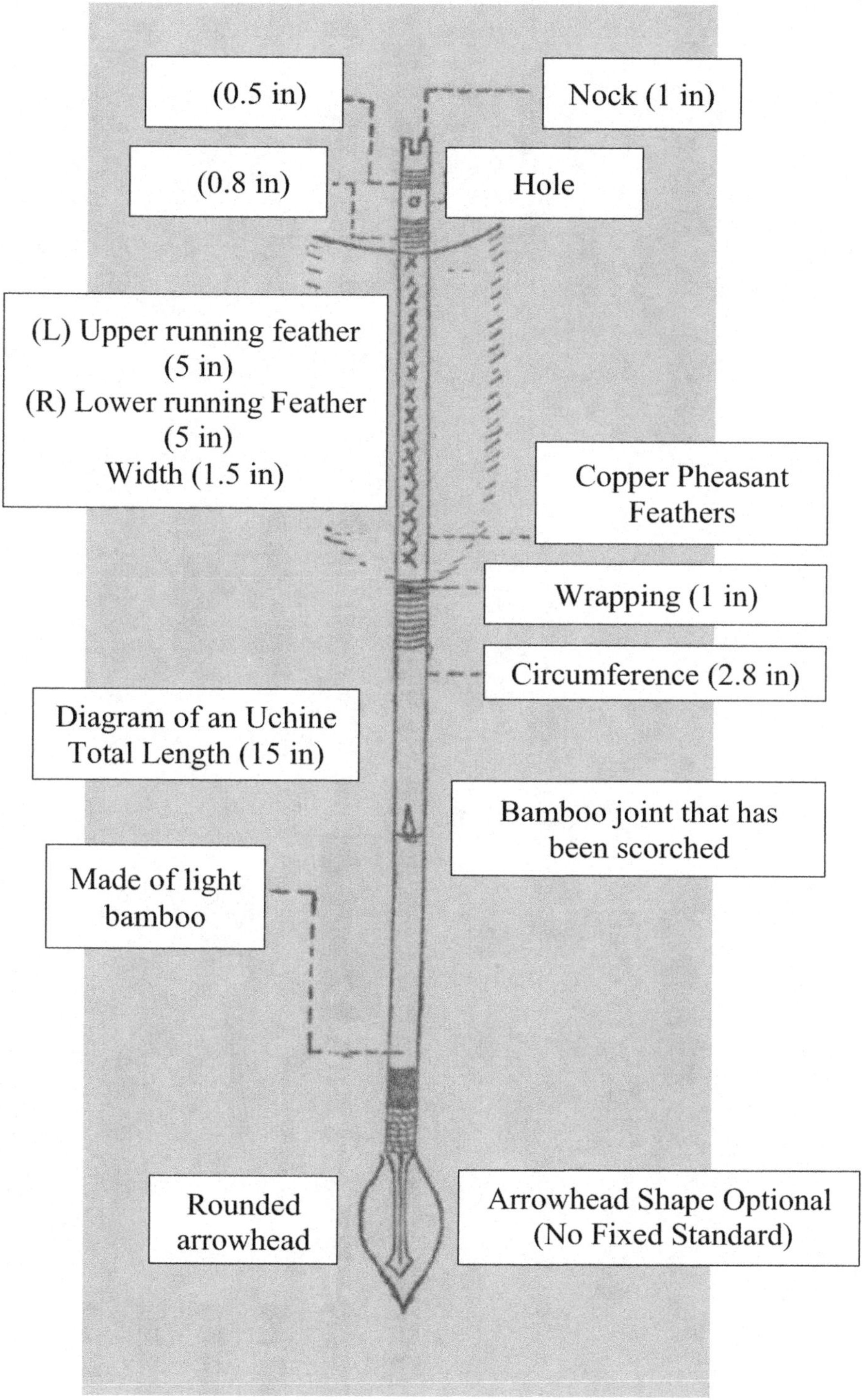
(0.5 in)
Nock (1 in)
(0.8 in)
Hole
(L) Upper running feather (5 in)
(R) Lower running Feather (5 in)
Width (1.5 in)
Copper Pheasant Feathers
Wrapping (1 in)
Circumference (2.8 in)
Diagram of an Uchine
Total Length (15 in)
Bamboo joint that has been scorched
Made of light bamboo
Rounded arrowhead
Arrowhead Shape Optional (No Fixed Standard)

手突鎗 （寫眞參照命正所有）

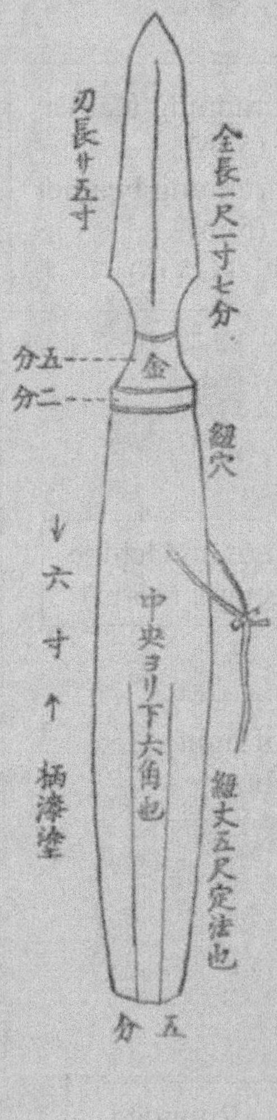

嘗て遊就館に出品し鑑定を乞ひたるる確然たる武器種類別を知らずと云ふ。

製作年代不明也漆色其の他を以て見るに五百年以前の者なるべし。

淺野秦次郎氏所藏品中

右手突鎗を見る構造形式略ぼ同じ柄に籐を巻きあり。

穂先柳葉形にして長さ三寸斗り柄の長さ七寸餘りあり。

二本對になり居れり（是れを以て見ると一手と云ひ度き感あり）

矢の部類打根なる乎。

一三一

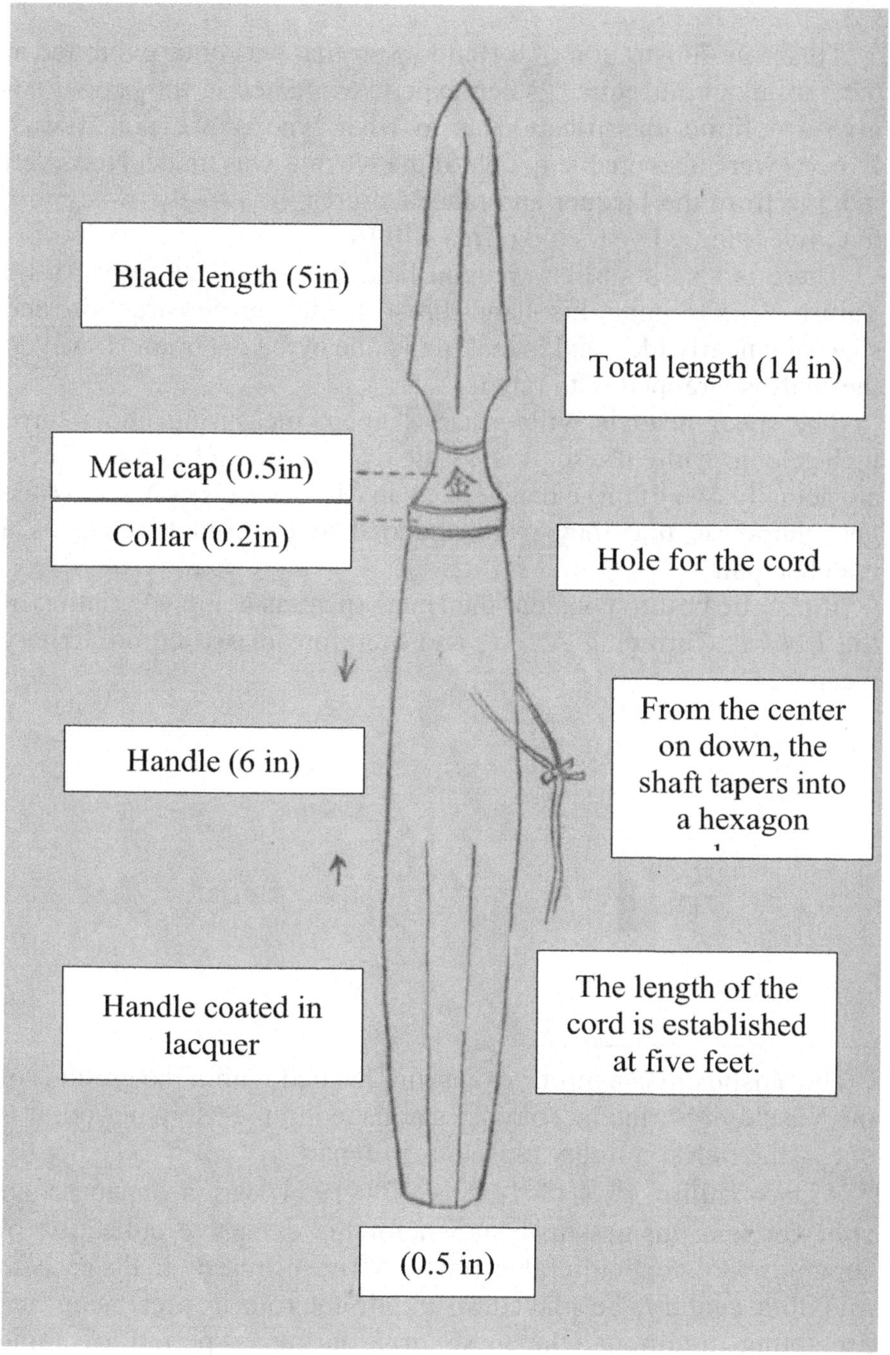
Blade length (5in)
Total length (14 in)
Metal cap (0.5in)
金
Collar (0.2in)
Hole for the cord
Handle (6 in)
From the center on down, the shaft tapers into a hexagon
Handle coated in lacquer
The length of the cord is established at five feet.
(0.5 in)

Teyari, Hand-spear

This is an illustration of a Hand-spear that was once exhibited at the Yushukan museum.[39] When experts evaluated it, they could not give a definite classification as to what type of weapon it was. Experts were also unable to determine when it was made. However, judging from the lacquer and other features, it is likely to be more than five hundred years old (circa 1400.)

There is also a similar weapon held in the collection of Asano Taijiro.[40] Examining his hand-thrust spear, its construction and shape are nearly identical to that of the one in the diagram. However, the shaft is wrapped with rattan.

The spear head is willow-leaf shaped, measuring about three inches long, while the shaft is a little over seven inches long. There are actually two of these hand-spears in Mr. Asano's collection (one gets the sense that they were intended to be used together as a matched pair.)

It may be best to consider the Hand-spear as being a variation of the *Uchine*, Throwing Arrow, and therefore classified an archery weapon.

[39] The Yushukan is a military museum located within the grounds of the Yasukuni Shrine in Tokyo. Established in the Meiji period, it is one of the oldest military museums in Japan.

[40] Asano Taijiro 浅野泰次郎 1874~1954 was a Japanese art collector and businessman known for his extensive collection of ceramics and other cultural artifacts. Active primarily in the early to mid-20th century, he played an important role in preserving and promoting traditional Japanese arts during a period of rapid modernization.

Ninja Tools

Fukuyama Matsui
福山松翠 (?~1967)

Excerpt From:
Kogaryu Ninjutsu: Sento, Choho, Sosa Hikkei
甲賀流忍術 : 戦闘・諜報・捜査必携
Handbook of Koga School Ninjutsu:
Combat, Intelligence, and Investigation
Published 1953

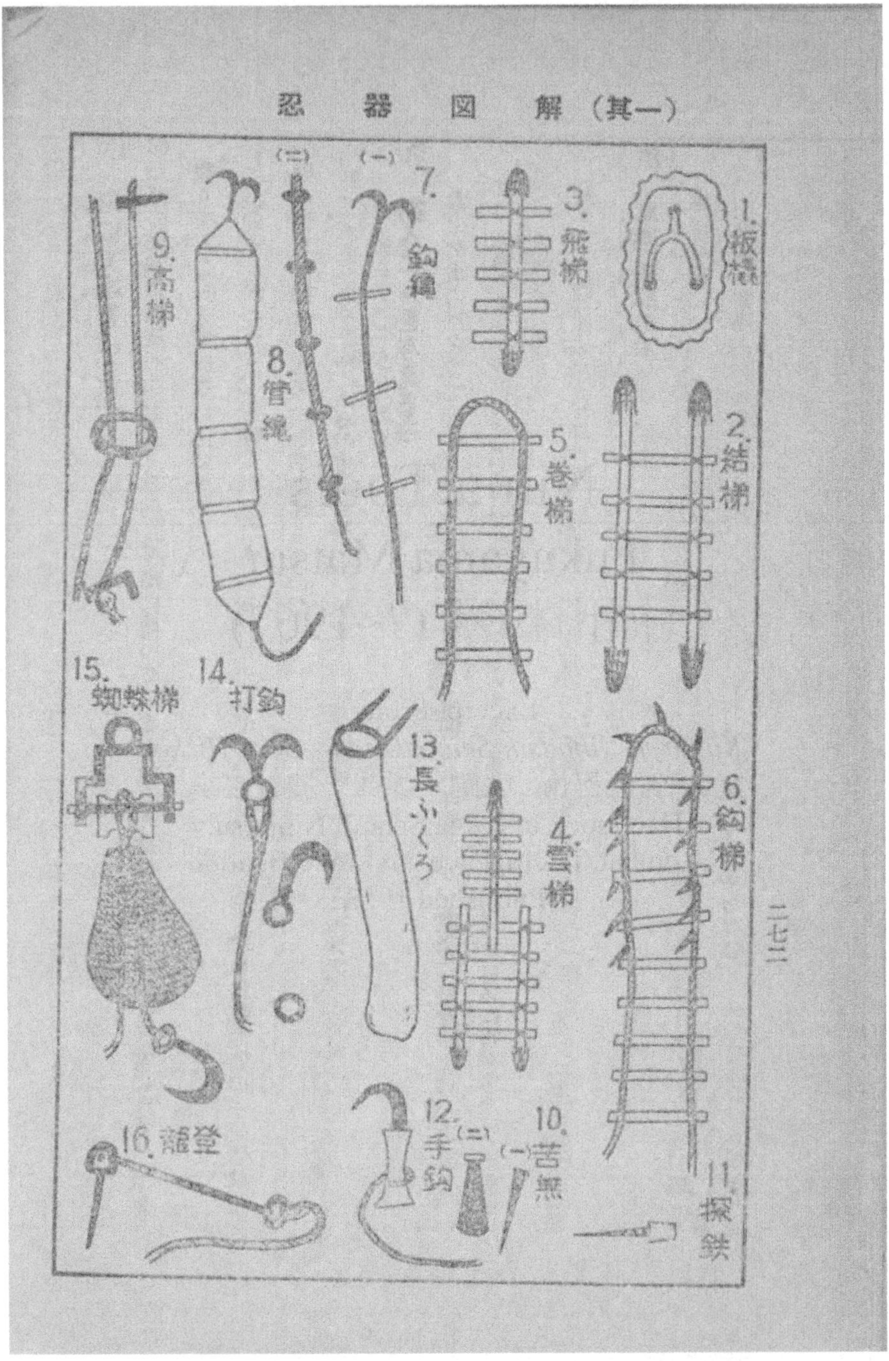
忍器図解（其一）
1.板橇
2.結梯
3.飛梯
4.雲梯
5.巻梯
6.鈎梯
7.鈎縄
（一）
（二）
8.管縄
9.高梯
10.苦無
11.探鉄
12.手鈎
13.長ふくろ
14.打鈎
15.蜘蛛梯
16.龍登
二七三

Ninki Zukai (Sono Ichi) 忍器図解 (其一) Ninja Tools Illustration (Part 1)[41]

1. Itabashi 板橋 – Board bridge / Plank bridge
2. Ketsujo 結梯 – Knotted ladder
3. Tobibashi 飛梯 – Flying ladder
4. Kusari-bashi 鎖梯 – Chain ladder
5. Makibashi 巻梯 – Rolled / Coiled ladder
6. Kusari-bashi 鉤梯 – Hook ladder
7. Kusari 鉤縄 – Hook rope
8. Tsunawa 管縄 – Tube rope / Pipe rope
9. Takabashi 高梯 – Tall ladder / High ladder
10. Kunai 苦黒 – Throwing spike
11. Tansetsu 探鉄 – Probing iron / Search spike
12. Tekagi 手鉤 – Hand hook
13. Nagafukuro 長ふくろ – Long sack / Long bag
14. Uchikagi 打鉤 – Striking hook
15. Kumo-bashi 蜘蛛梯 – Spider ladder
16. Ryuto 龍登 – Dragon climber

[41] Illustrations from:

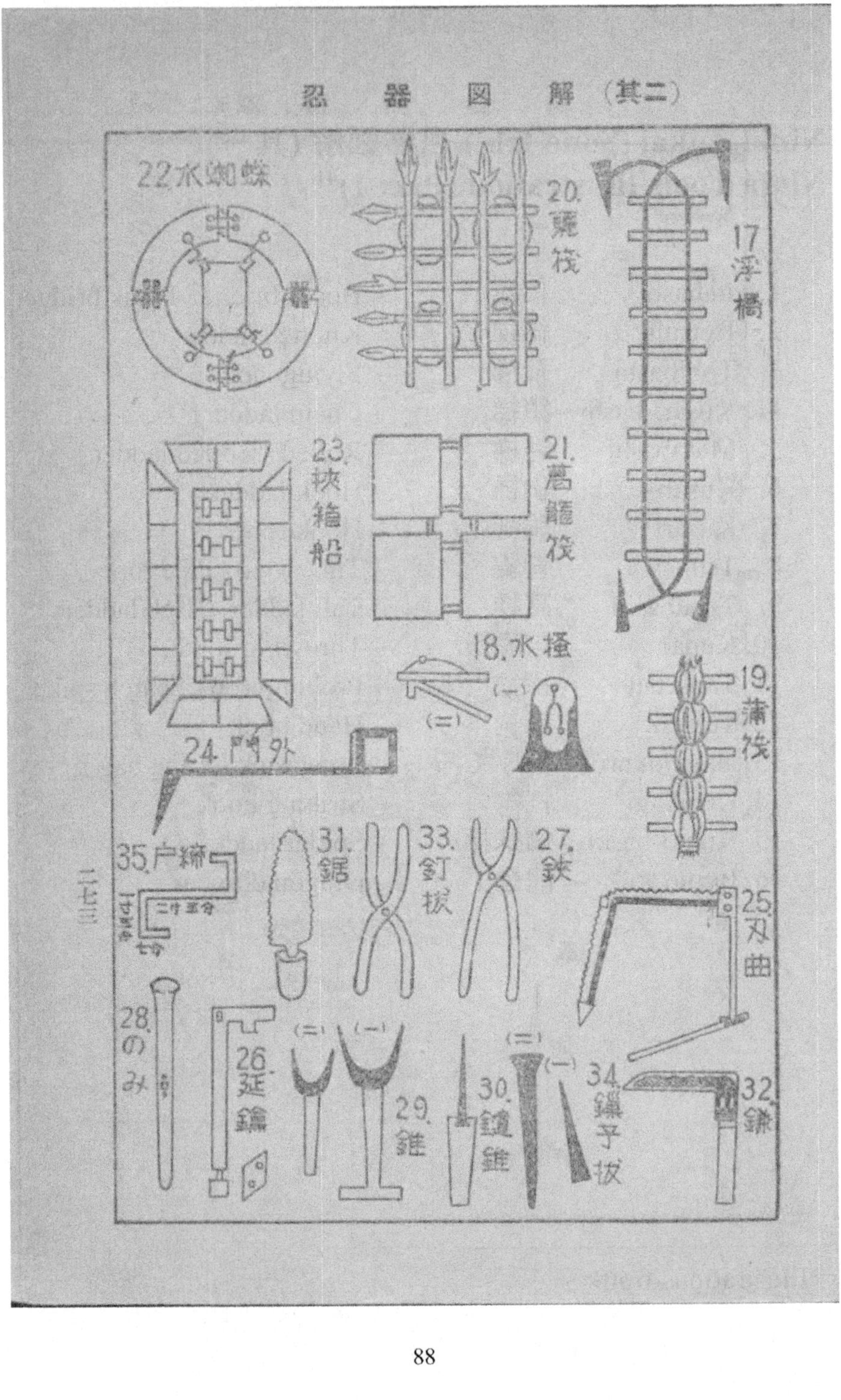
忍器図解（其二）
17 浮橋
18. 水掻
(一)
(二)
19. 蒲筏
20. 甕筏
21. 葛籠筏
22. 水蜘蛛
23. 挟箱船
24. 問外
25. 刃曲
26. 延鑰
27. 鋏
28. のみ
29. 錐
(一)
(二)
30. 鐶錐
(一)
(二)
31. 鋸
32. 鎌
33. 釘抜
34. 鑷子抜
35. 戸締
二七三

Ninki Zukai (Sono Ni) 忍器図解 (其二)
Ninja Tools Illustration (Part 2)

17. Uki-bashi	浮橋	– Floating bridge
18. Mizukagi	水掻	– Water rake / Water claw
19. Shiba-bashi	蒲梯	– Rush / Reed ladder
20. Kagibashi	篝梯	– Brazier ladder / Signal ladder
21. Takabisha	高鼈	– Tall turtle shell (climbing aid)
22. Mizugumo	水蜘蛛	– Water spider (water-walking)
23. Kakubune	挾箱船	– Box boat / Portable boat
24. Mon-gai	門外	– Outside the gate
25. Enkyoku	円曲	– Circular bend / Curved saw
26. Nobin	延鑢	– Extendable file
27. Tetsu	鉄	– Iron (general iron tool)
28. No-mi	のみ	– Chisel
29. Kui	錐	– Awl / Drill
30. Kusari-kui	鎖錐	– Chain awl
31. Nokogiri	鋸	– Saw
32. Kama	鎌	– Sickle
33. Kuginuki	釘抜	– Nail puller
34. Kamae	鎌子	– Small sickle
35. Tobiso	戸締	– Door lock / Bar lock

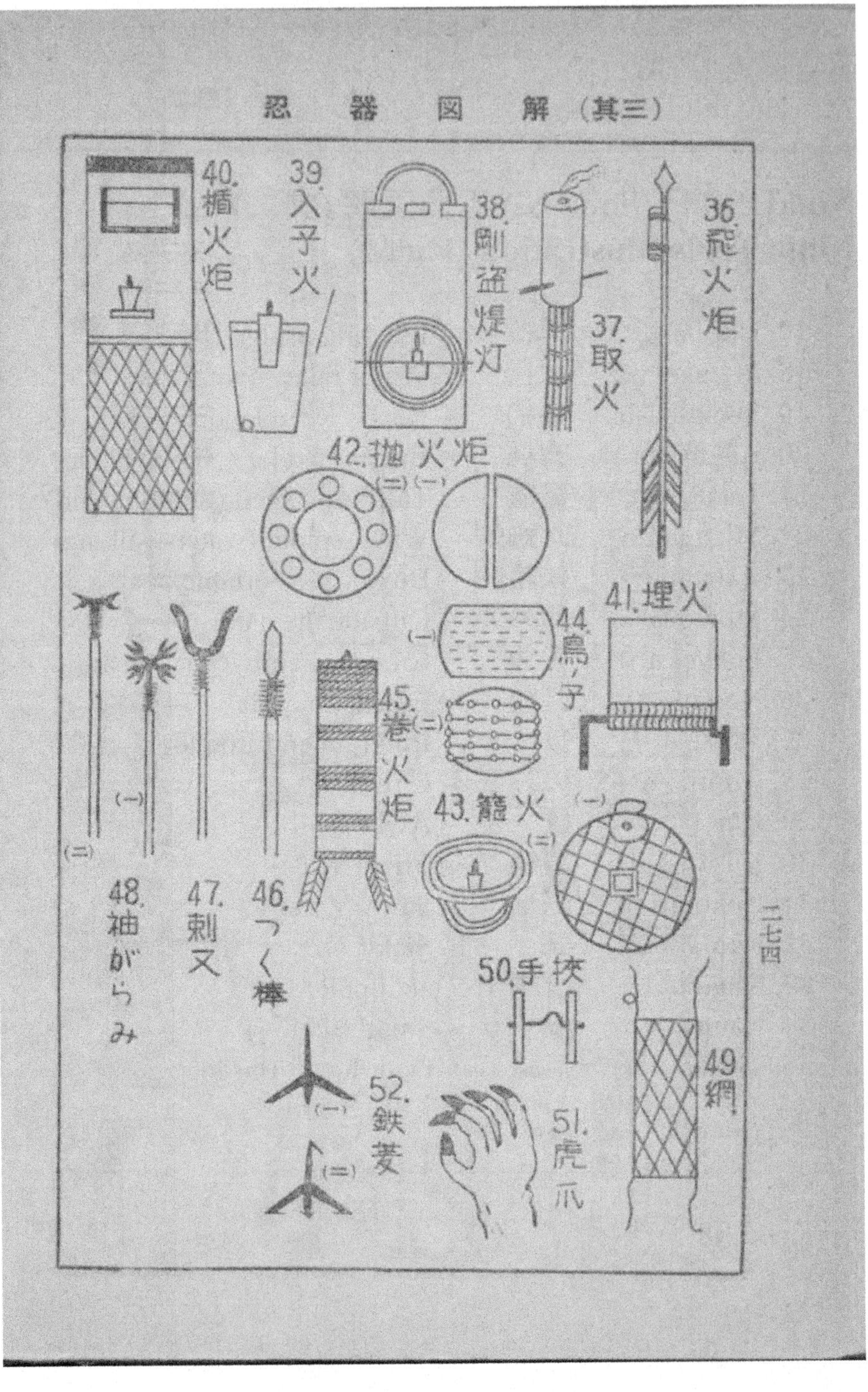
忍器図解（其三）
36.飛火炬
37.取火
38.剛盗提灯
39.入子火
40.楯火炬
41.埋火
42.抛火炬
(一)
(二)
43.籠火
(一)
(二)
44.鳥子
(一)
(二)
45.巻火炬
46.つく棒
47.刺又
(一)
(二)
48.袖がらみ
49.網
50.手挟
51.虎爪
52.鉄菱
(一)
(二)
二七四

Ninki Zukai (Sono San) 忍器図解 (其三)
Ninja Tools Illustration (Part 3)

36. Hibihijo 飛火炬 – Flying fire torch (arrow)
37. Toribi 取火 – Fire starter
38. Goto renjo 剛盗煉灯 – Burglar's lantern
39. Iriko hi 入子火 – Inserted child fire
40. Tate hijo 楯火炬 – Shield fire torch
41. Umebi 埋火 – Buried fire
42. Nage hijo 抛火炬 – Throwable fire torch
43. Kago hi 籠火 – Basket fire
44. Tori ko 鳥子 – Bird child
45. Maki hijo 巻火炬 – Rolled fire torch
46. Tsukubo つく棒 – Striking staff
47. Sasumata 刺又 – Restraining pole
48. Sode garami 袖絡み – Sleeve entangler (two types)
49. Ami 網 – Net
50. Tekagi 手挟 – Handcuffs
51. Tora no tsume 虎爪 – Tiger claws
52. Tetsubishi 鉄菱 – Iron water caltrops

Secrets of Ninjutsu

1776

Author Unknown

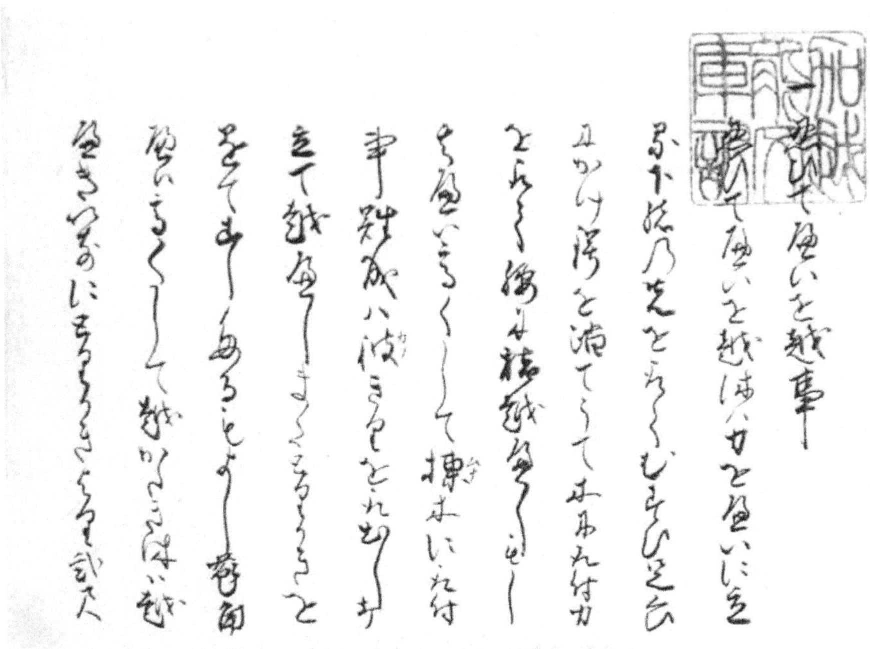

忍びてへいを越事
忍びてへいを越つ八力（刀）をへいに立
る、下緒の先を取てむすび足くび
にかけ鍔を踏でうで木に取付、力
を取て腰に指越べしもし
其へい高くして棟木に取付
事難成八彼きりを取し打
立て越べし、またとりがきを
懸てこしたるもよし、兎角
へい高くして越がたき体八越
べき以前にとりがきより弐尺

Illustration from *The Essence of Ninjutsu* by Gingetsu Itoh 1917

How to Stealthily Climb Over a Fence

To climb over a fence stealthily, set your sword upright against the fence. Take the end of the Sageo cord, tie it and loop it around your ankle. Next, step on the Tsuba, hand guard, as a foothold, reach up and grasp the crossbeam. Then pull your sword up and slide it back into your belt. This is how to climb over a fence.

If the fence is still too high to reach the crossbeam, take a nearby branch from tree, drive it firmly into the ground, and use it to climb over.

Alternatively, you can also use a portable ladder with hooks on the end, and scale the fence that way.

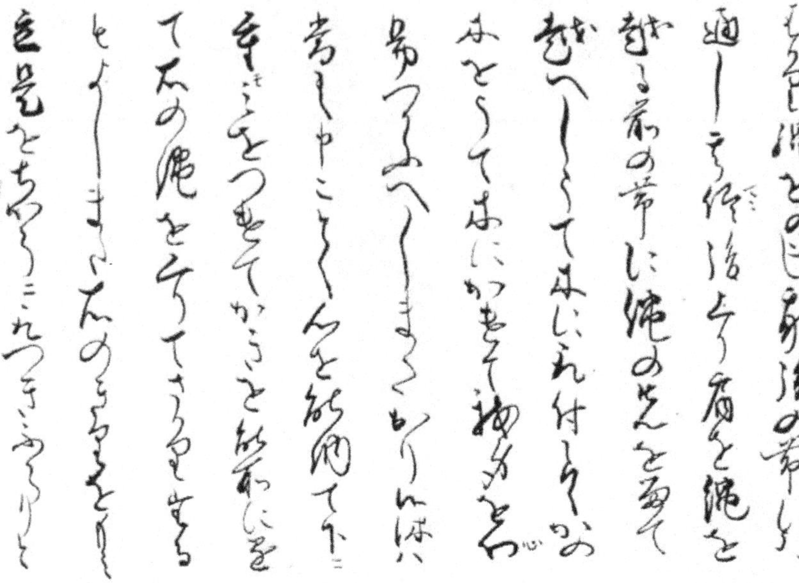

はかり縄をのこし我後の帯に引
通し其侭後上り肩を縄を
越る前の帯に縄の先を留て
越へし、うで木に取付候ハバかの
木をうで木にかけて我身を心
易つかふべし、またおり候体ハ
常々申ごとく心を能納て下ニ
重モミをつけてかきを能所に懸
て右の縄をくりてさがりたる
もよし、右のきりをもミ
立是をちから二取つきぶらりと

Scaling a Tall Fence

In any case, if you are faced with scaling a tall and difficult to climb fence, measure two Shaku, two feet, from the bottom of your *Kagi Nawa*, rope and hook. Tie that section to the back of your belt so two feet remain hanging down. Just as you are at the top of the fence, pass the rope tied to your belt over your shoulder, around the crossbeam and tie it to the front of your Obi. This will allow you to easily cross over the top of the fence.

Descending

Next, regarding descending. As has been said before, be sure to focus your mind. Regarding descending, the following method is recommended. As has been said repeatedly, focus and calm your mind before shifting your weight downward. Be sure your hook is set firmly, then, while adjusting the rope with your right hand, descend the wall.

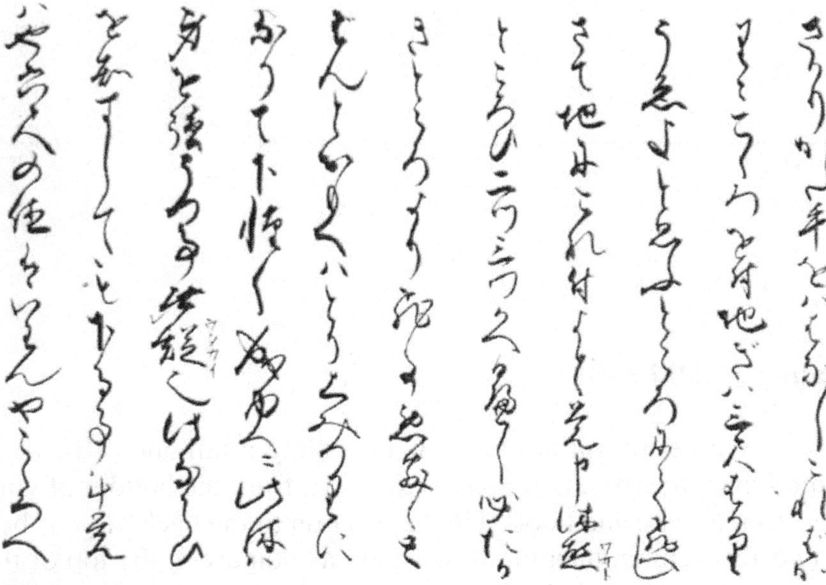

さがりかた手を八なし、こればかり二ここ
ろを付地ぎ八三尺ばかり
うえと思ふところにて飛べし
さて、地にこれ付よと覚申休態
ところび二ツ三ツるべし、必たか
きところより飛事悪敷くと
ぞんとおもへバとり上みつりに
ふりて下軽く成ゆへ二此体八
身を強うつ事無疑也、此ならひ
を知ずしても下る事計覚
ばや、六尺の値はいわんやこころへ

Ensure your hook is set firmly and then, holding onto the hook with both hands begin to lower yourself. Let go with one hand and when you judge yourself to be about three Shaku, three feet, from the ground, drop down.

Regarding Landing

Next, is an important lesson on how to land. The moment you make contact with the ground, be sure to roll two or three times. Most people think that jumping from a high place is dangerous, however if you add a swing to the rope before you drop, your body will become lighter the moment you make contact with the ground. Using this method means you won't strike the ground hard.

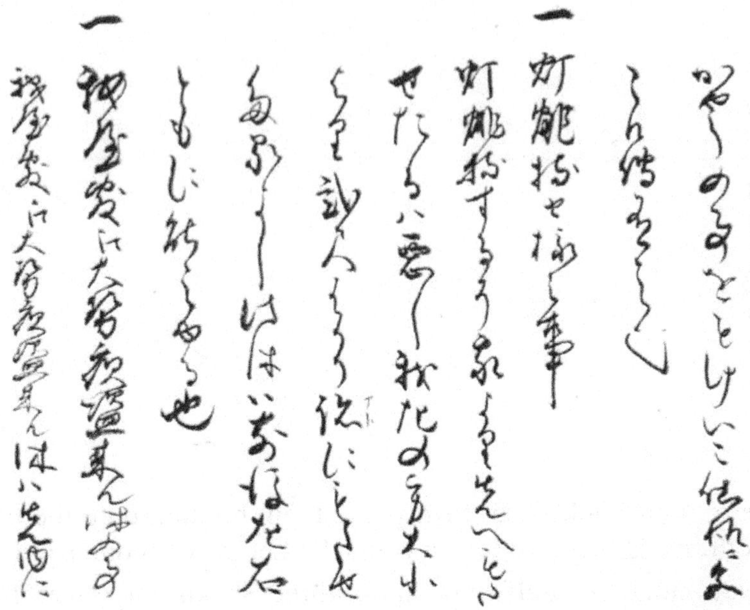

たる人ガ身をうつことすくなし
かやうの事をも、けいこ仕様二色
々候伝有也
一　提灯持セ様之事　提灯持するに我より
先へもた　せたる八悪し我左の方大小よ
り弐尺ばかり跡にもたせ　たるよし、この
体八前後左右　ともに能ミゆつ也
一　我屋敷江大勢夜盗来ん体の事　我屋敷
江大勢夜盗来ん体八先内に

Even if you do not know this technique of dropping to the ground, it is best to at least to learn how to drop from a height. Needless to say, this is in regard to dropping distances of more than six Shaku, six feet. Those who have trained this method rarely strike the ground hard. There are many different ways of training this method.

Walking With a Lantern Bearer

As for how to have a lantern carried, it is not good to position your lantern bearer ahead of you. Instead, have the bearer walk on your left about two Shaku, two feet, behind the end of the scabbards of your swords. This will allow you to see clearly in all directions, front back left and right.

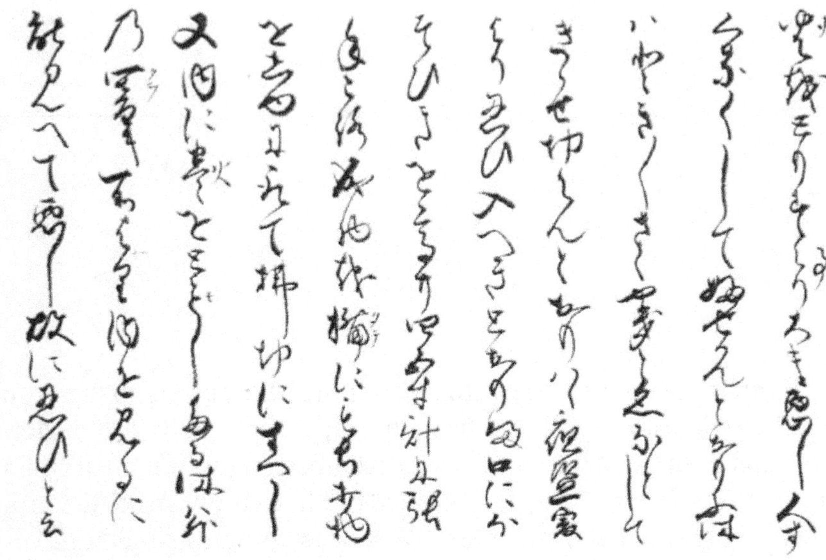

火をともす事、大キニ悪し、人す
くなくして、ふせがんとたもふ体
ハときどきささやきこえなどして
きかせ切らんとおもハバ、夜盗爰
より忍び居るへきとあり、多口に分
そひきを高サ四五寸計に張
手ごろ成物を盾にもち打物
を口口取て拂切にすべし
又打ちに火をともしたる体八外
の闇キ所より内を見るに
能見へ悪し故に忍びと云

Defending Your Household Against Thieves

If you realize that a large number of *Yato*, night thieves, are attempting to enter your residence, it is extremely unwise to light lamps.

If there are only a few people in your household, you can make the intruders think, “There are many people in this room!” by intermittently make whispering sounds.

In addition, tie tripwire cords across doorways and various spaces about 4~5 Sun, 4 inches, high.

Next, find a suitable object to use as a shield, and locate an object that can be used as a striking weapon. Attack the thieves with sweeping strikes.

It is important to remember that if you have lanterns lit inside your house, prowlers outside in the darkness can clearly see within, which is disadvantageous. Thus remember the importance of Ninja, or stealth and concealment.

There is a Kuden, an orally transmitted teaching.

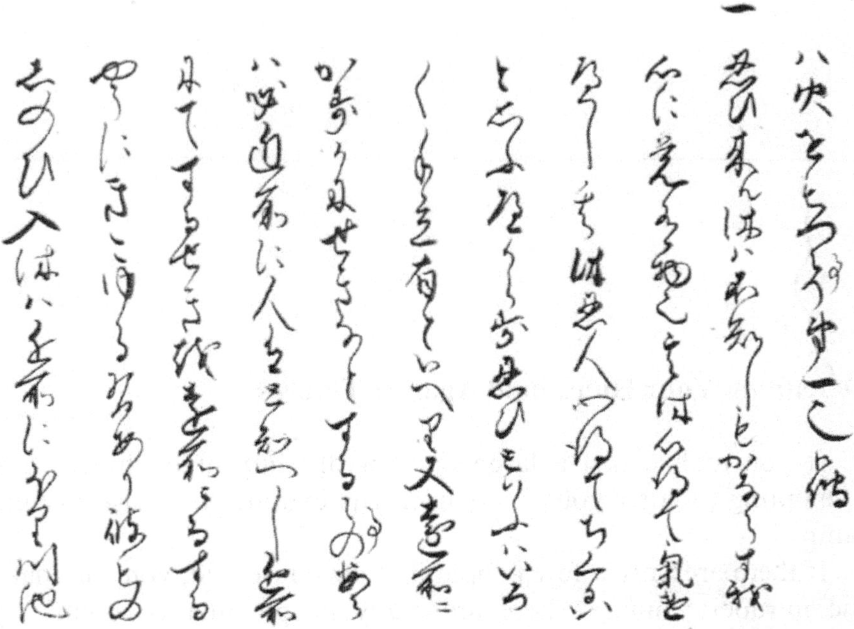

八火をもつ事第一也口伝
一忍び来ル体八不知しもかならず、我
心に覚有物也、其体心得て気遣 べし、
其体忍人心得てちくるい
と思ふべからず、忍びといふ八いろ い
ろ手立有といへり、又遠所二 かすかに
せきなどする事あら
バ必近所に人有と知へし、近所
にてするせきを遠所二てする
やうにきこゆる習あり、彼もの
しのび入り体八近所にほり、川、池

Kuden

Thus, when lights are burning in a house, the interior is easily visible in from the darkness outside, which is disadvantageous. Therefore it is of primary importance for a Shinobi, a Ninja, to not carry or use fire.

Sensing a Ninja

If you ever sense that a Ninja is near, even if you cannot clearly define what you feel, understand that your mind has perceived something and you should pay heed to that. In such a case, you should remain vigilant.

If you should sense something, you may not realize it is due to a Ninja being near, thus you should never dismiss it as something out of your imagination. Remember that Ninja have many methods at their disposal for achieving their goals.

For example, if you were to hear something off in the distance that sounds like someone suppressing a cough, this should make you realize that someone is nearby. This means that though it may seem the sound is distant, a Ninja is actually making the sound at close range and only making it seem as if it were distant.

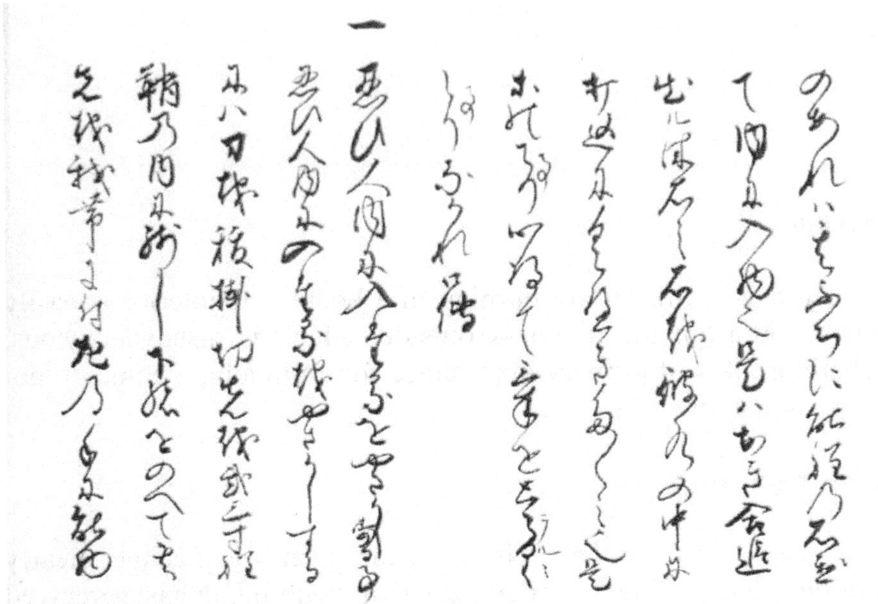

のあれバ、其ふちに能程の石置 て内
に入物也、是八出き合追 出ル体右之
石を彼水の中に 打込に具べきたくミ
也、是 等の事心得て気を□□るる
事なかれ 口伝
一忍び人内に入たるをやさがしする
事 忍び人内に入たるをやさがしする
に八刀を抜掛切先を弐三寸程 鞘の内
に残し下緒をのへて其 先
を我帯に付左の手に能物

Also, before a Ninja enters a moat, river, or pond they will place a large stone at the edge. This is used to create a diversion in case they are discovered. If being pursued the Ninja can throw the stone into the water to draw attention to that spot.

You should take to heart the lesson of never letting your guard down.

There is a Kuden, orally transmitted secret.

An Eishin School Document showing this technique
無双直伝英信流居合兵法

How to Find a Ninja in Your Home

This explains how to search for a Ninja who has snuck into your house.

When searching, use your sword. Draw the blade out of the scabbard, leaving just 2~3 Sun, 2~3 inches, of the tip inside. Pull the Sago cord free and hook loop it through your belt. Hold your sword in your left hand.

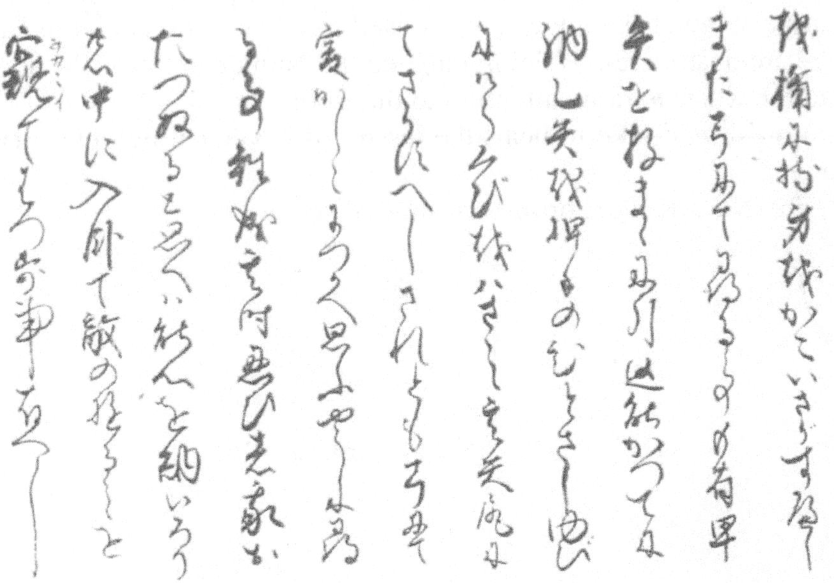

あとは自分で考えてくださいを盾に持身
をかこいさがすべし また弓にて尋る事も
有、早 矢を存ままに引込能かつてに 納口
矢を押手のひとさしゆび にけらくび螻蛄
首（けらくび）をハサミ其矢尻に てさが
すべし、されども弓にて 爰かしこにつか
へ思ふやうに尋 る事難成、其時忍び者我
出 たずぬると思へバ能心を納おいろり の
中に入臥て敵のゆるミを 窺てはづす事有
べし

Holding the Haya and Otoya arrows
大和流弓道伝書 Yamato School Archery Manual
Murakawa Kiyoshi 村河清 (?~?) 1926

There is also a method of searching with a bow. Take your bow and one set of arrows, a Haya, former arrow, and a Otoya, latter arrow. Place the Haya on the bowstring and draw back as far as you can with your right arm. Hold the *Kera-buki*, or nock of the Otoya with the index finger of your left hand, which is also holding the bow. Use the arrowhead of the Otoya to search the room.

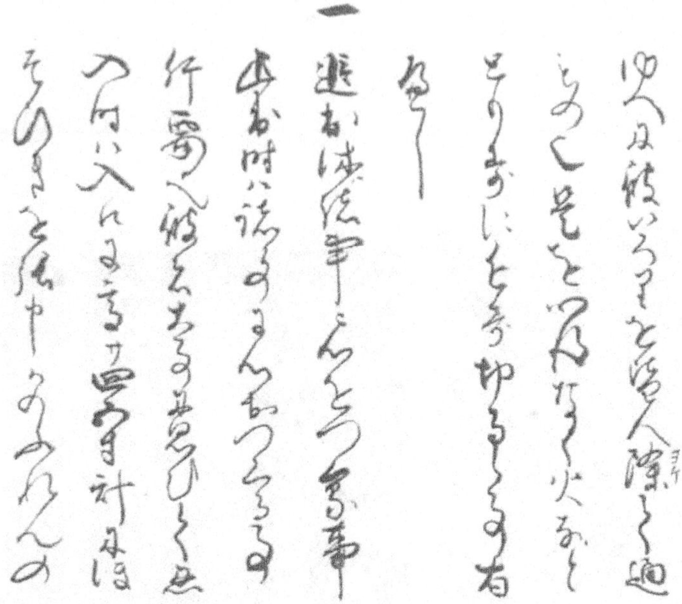

扨いろりの中に身をかくす事
起合候えども常々見置たる
ゆへに彼いろりを皆人除て通
もの也、是を心得なく火など
ともすに近寄切るる事有
べし
一追出体諸事ニ心をつくる事
追出時八諸事に心おつくる
事 肝要也、彼者大事に思ひて忍
入時八入口に高サ四五寸計ほ
そひきを張申か、のふれんの

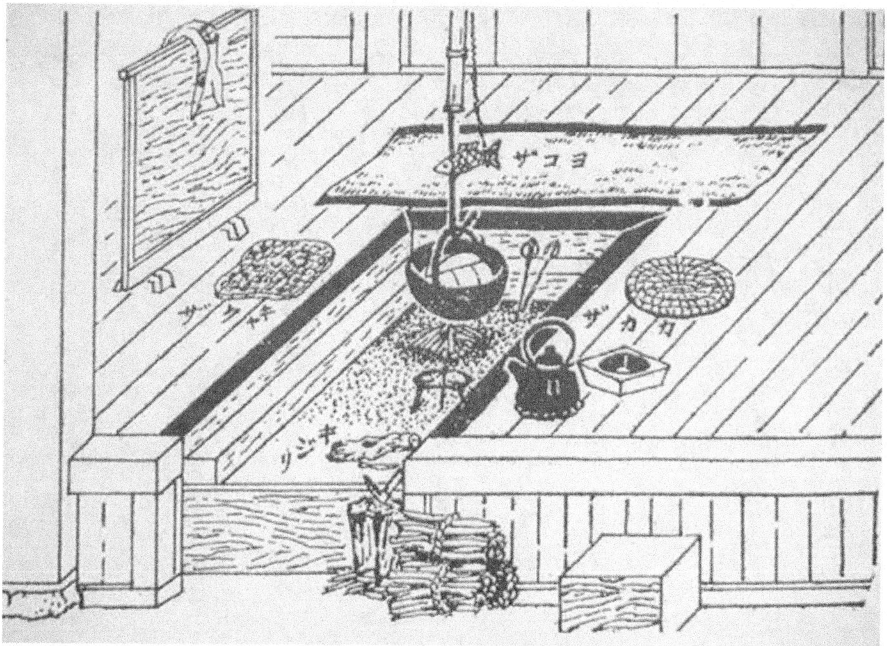

Irori hearth in a typical house

Minzokugaku Jiten 民俗学辞典 *Dictionary of Folklore*
Yanagita Kunio 柳田國男 (1875–1962)

However, using a bow to search around your house is quite difficult and likely will not go as your intend. At such times, if the Ninja believes that you have left the room, he will quietly compose himself, hide within the sunken hearth common to houses and wait until he has determined you have relaxed. The Ninja will then seize that opportunity to slip away.

Ninja Will Hide in the Hearth

Hiding inside the Irori hearth is a common technique. The reason this spot is used is because people are accustomed to seeing it in everyday life, so they tend to avoid passing directly over it. However, if you are not aware when approaching the hearth, say to light the cooking fire, you may be cut down.

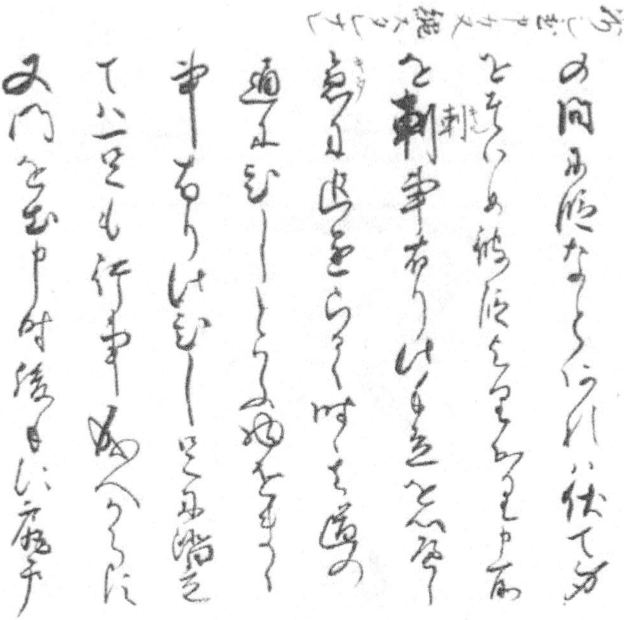

あれバ上より下へすだれのごと
く切さき〇八左右のはつれ取て
〇下四五寸計残シ置申カ、又縄スダレナレ
むすびをく事も有　扨次々
の間に段などあれバ伏て身
をそい如彼段よりおり申所
を斬事有り、此手立を心べし
急に追懸けらるる時、其道の
通にひしという物まく 事有り、
此ひし足に踏立 てバ一足も行事成べから
ず 又、門を出申時、後手に尻戸

Street scene showing shops with long Noren curtains. Edo Era print

Considerations When Trying to Expel a Ninja

There are many things to be extremely careful of when trying to expel an enemy from your house. If you own something that a Ninja wants, he will consider it essential to obtain it. Prepare by stretching a thin cord across the entrance of your house about 4~5 Sun, 4~5 inches, off the ground.

If you have a Noren curtain hanging in front of your door, make vertical cuts in the cloth. You can also cut the cloth horizontally and tie the ends together.

○ You can also hang four or five inch lengths of cord from the bottom of the Noren curtain.[42]

[42] This is the small horizontal text at the top of the page.

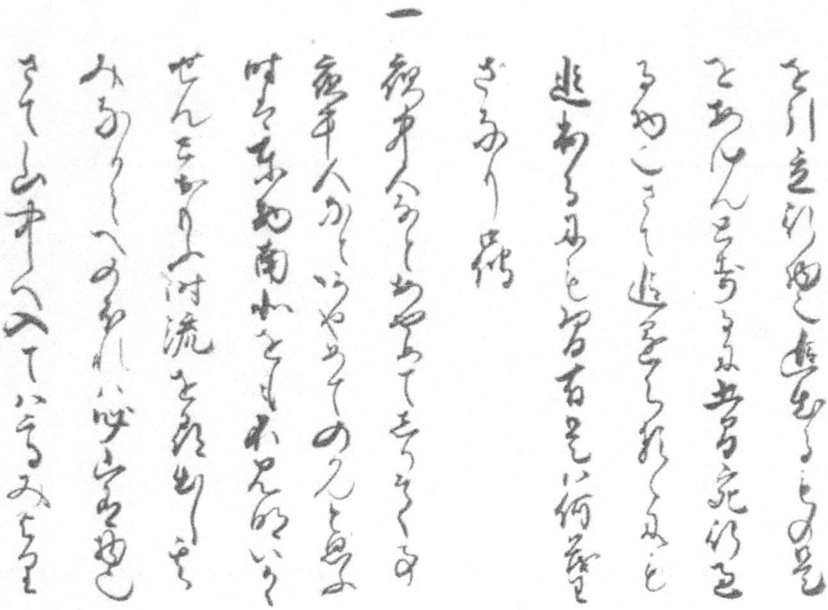

を引立行物也、追出るもの是ここは尻戸
を安心んとするに五間宛行過る
物也、さて追懸らるるにも
追出るにも間有、是八何もわ
ざなり　口伝
一夜中人などあやめてしりぞく事
夜中人などあやめてのかと思ふ
時は東西南北をも不見明いかが
せんとおもふ時、流を尋出し、其
みなかミへのぼれバ必山有物也
さて山中へ入て八高みより

Killing a Person at Night and Escaping

At night, after you have killed someone and you are trying to orient yourself, use this method to determine the proper direction. To find east, west, north, or south, first seek out a stream and move upstream. This will invariably lead you to the mountains.

Having entered the mountains, find the highest point you can and look about for the glow of Taimatsu torches, which will indicate pursuit. You will also be able to determine where the paths up the mountain are by their movements.

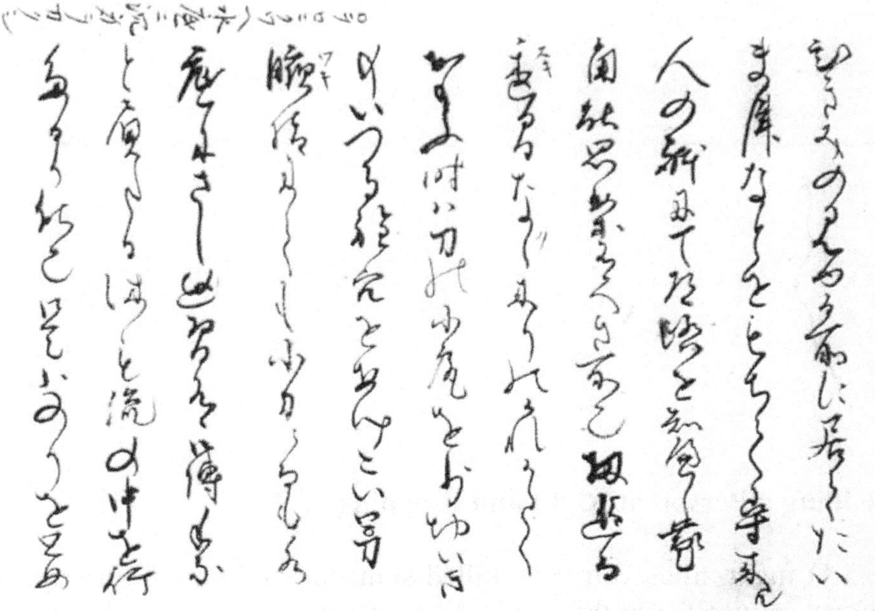

ひきみの見やる所に居てたい
まつなとをもちて尋来ル 人の体にて道
路を知べし兎 角能思案有べき所也、扨
追而 透間なく来たり、 のがれがたくぉ
もふ時八刀の小尻を少切いき のいつる
程穴をあけこい口〇刀
〇ヲ口にクツへ水底ニ沈身ヲカクシ 脇
指 にても小刀ニ而も水 底にさし込習有
口伝手な ど負たる体も流の中を行 た給
う時也、 是八のりをとめ

In this case you need to consider the situation carefully before moving. If pursuit is close on your heels and it seems doubtful that you will be able to outrun them, draw your sword, cut off the end of your scabbard to make a breathing hole. Put your mouth on the opening of your scabbard, called the *Koiguchi* carp's mouth, and submerge yourself in water to hide.

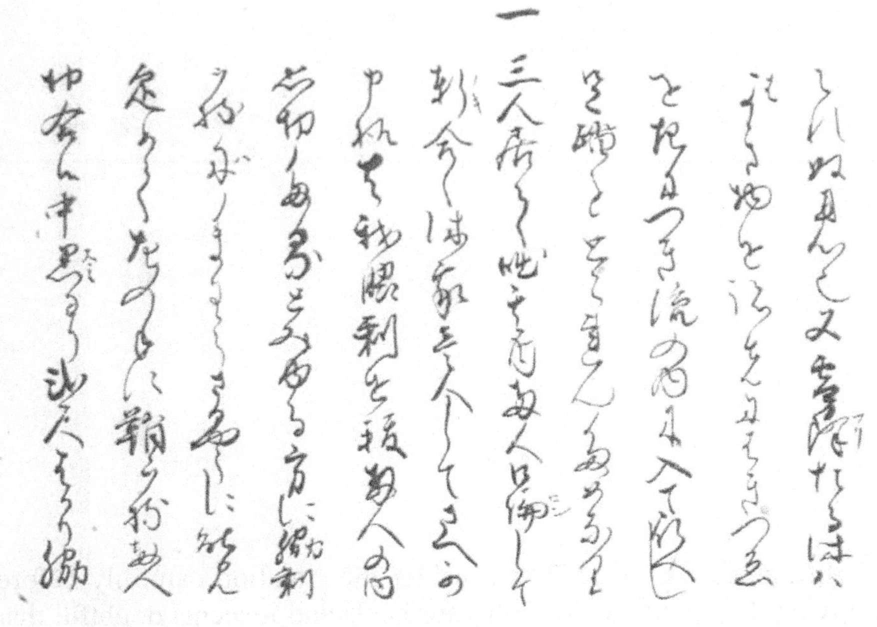

られぬ用心也、又雪降りたる時ハ
はき物を跡先にはきつえ
を左につき流の内に入て行べし
足踏をとられんためなり
一　三人居て咄其内両人口論して
斬合の時、我壱人してさへか
申様者、我脇刺を抜両人の内
思切たるとみゆる方に脇刺
ヲ持ハノまわらざるに能見
定めて左の手に鞘ヲ持両人
切合候中墨より弐尺ばかり脇

Breaking up a Sword Fight

If there are three of you and the other two begin an argument that devolves into a swordfight, here is what to do if you decide to involve yourself. Draw your sword and face the man who appears more determined to see the fight to the end.

While carefully observe the movement of their blades, also pull your scabbard out of your belt and hold it in your left hand while your sword is in your right.

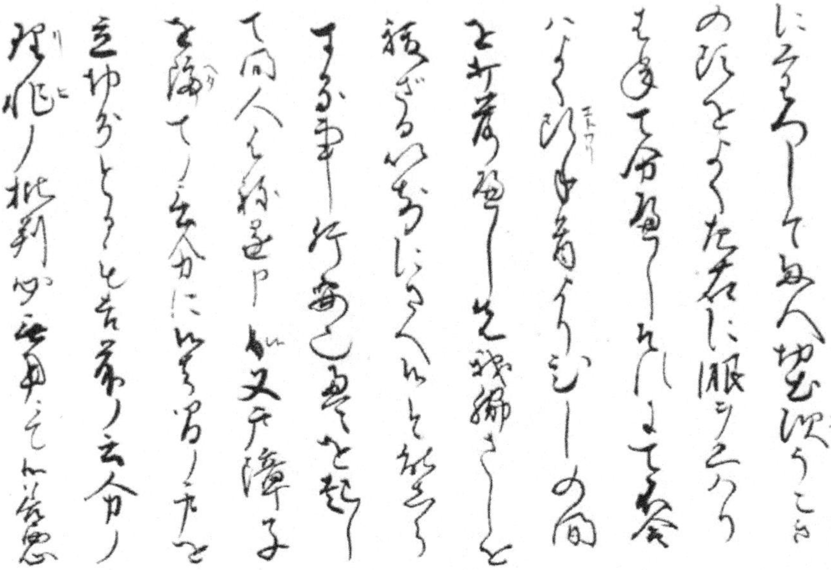

にくわつして両人切出すうごき
頭をよく左右に眼ヲくばり
はねて分べし、先我脇さしを 抜ざる
以前にさへ候と能しら する事肝要
也、畳を起し て間人はね懸申候、又
戸障子 を隔てノ言分に候者間ノ戸を
立切分とるるも吉、前ノ云分ノ 理非
ノ批判必無用ニて候、善悪

Step to within about two paces of the combatants on an angle with your sword directed towards the man with greater resolve. The moment they seem about to strike, leap in while keeping both men in your field of vision and disrupt the duel.

Even before you draw your sword, you should not project any intent. This is an essential point.

You can also break up a duel by pulling a Tatami mat up and throwing it between them.

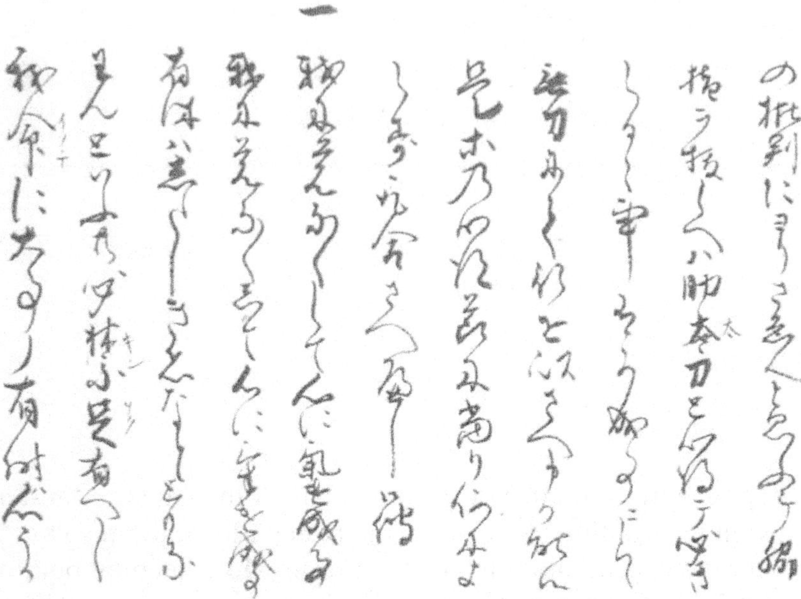

の批判にヨリさえんと思ふエオ脇
指ヲ抜候へバ助太刀と心得テ必き
らるる事有可成事ニ候ハバ
無力にて行を以さへ申が能候 是等の心得
節に当り何によ らず取合さへべし 口伝
一 我に覚なくして心に気遣成事
我に覚なくして心に気遣成事
有時ハしたしき者などともな わんといふ
共、必禁足有べし
我命に大事ノ有時ハ心うか

Stopping an Argument

If two people are arguing across a wooden door or paper door, it is best to just insert another door between them in order to separate them. You do not need to concern yourself about who is right or wrong in the argument.

On the other hand, if you decide to intervene on one side or the other and draw your Wakizashi sword, this action will be seen as designating yourself *Tsukedachi,* a backup swordfighter. Doing such a thing may result in you yourself being cut down. In such a case, it is better to act without force and restrain them through one's conduct.

These are important points to consider and you should base your decision on the satiation at the time. It is essential that you separate the two parties involved.

Trust Your Instincts

Sometimes, for no apparent reason, you may feel a sense of unease. If so, even if a close acquaintance were to invite you to accompany them, refrain from going out.

Invariably, when there is a something that will have great consequences on your life, you will not feel at ease. If such a feeling comes over you, it is best to measure your pulse.

The way to do this is to use your right hand to check your pulse at the throat. Then check your pulse on your right wrist with your left hand and compare the two. If the rhythm of your pulse is the same at both places, there is no problem.

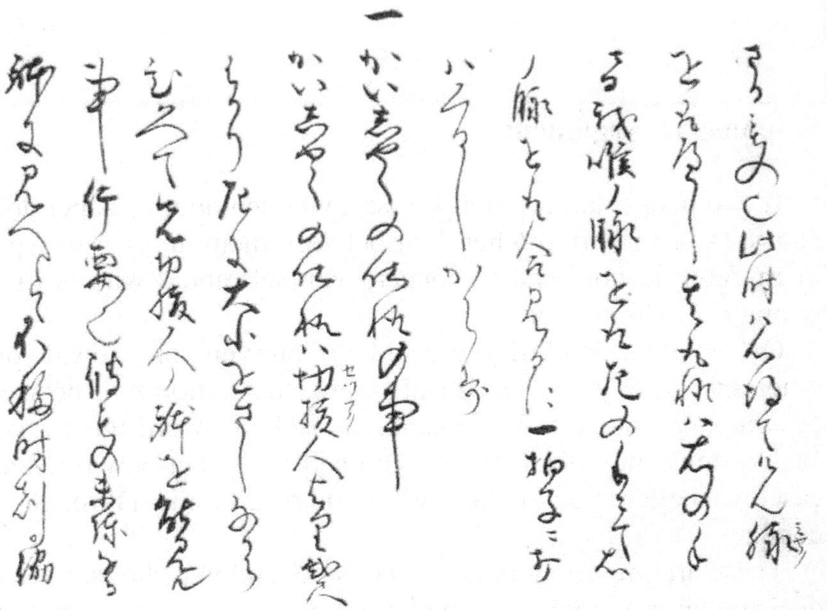

ざるもの也、此時ハ心得てけん脈
を取べし、其取様ハ右の手
二而我喉ノ脈を取左の手二て右
の脈を取合見るに一拍子二打
バくるしからず
一　かいしゃくの仕様の事
かいしやくの仕様ハ切腹人より弐尺
ばかり左に大小をさしながら
ひかへて先切腹人ノ躰を能見ル
事肝要也、彼もの未練なる
体に見へ候ハバ　不移時刻〇脇

Seppuku scenes from The Forty-Seven Ronin
Edo Era scroll

Role of the Second in Seppuku

The Kaishaku, or second, is the one who assists the person sentenced to Seppuku. You should stand with both long and short swords in your belt about two Shaku, two feet, to the left of the person committing Seppuku. It is essential that you carefully observe the condemned's attitude and movements.

If the Condemned is Unprepared to Follow Through

If the condemned appears to be appears hesitant or distressed, you should, without delay, draw your sword and cut. Wait only for the moment his neck is at a good angle and not moving to cut.

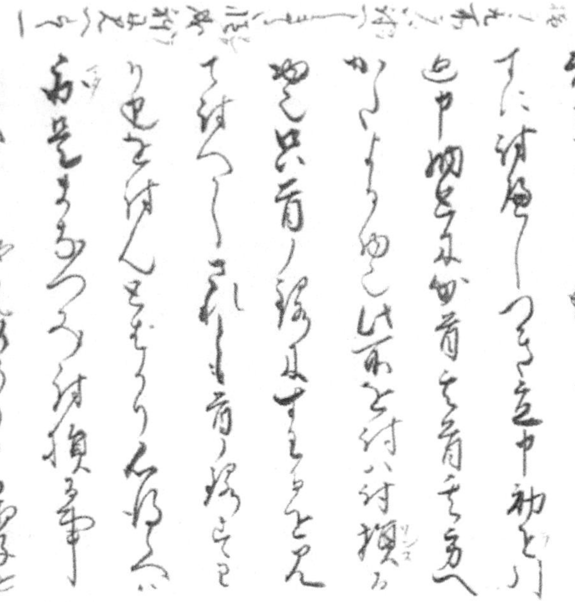

○脇指ヲ取所ヲバ討べし、また程成体
に見へ候ハバ 一刺引廻首ノ碌にすわる
をあい ずに討べし、つき立申初を引
廻申胴とに分首其首其方へ
かたよる物也、此所を討ハ討損る
や也、只首ノ碌にすわるを見 て討へ
し、されども首の碌すわ り申也を討ん
とばかり心得候へバ 亦是まなつみ討損
る事 有べし我気の、のりたる拍子を
討べし、前に切腹人ノみぬやう二刀を
ぬきて取合い勝手ノ能所に

Timing Your Cut

Following the initial thrust of the knife by the condemned, his head and neck will extend away from an upright sitting position, inclining to one side or the other. If you do not wait for that moment, your cut will likely fail to sever the head.

However, be aware that you cannot simply wait for the condemned's head to stop moving before cutting. You must focus on the proper timing and failing do so will result in a poor cut. You should cut in a rhythm aligned with your spirit.

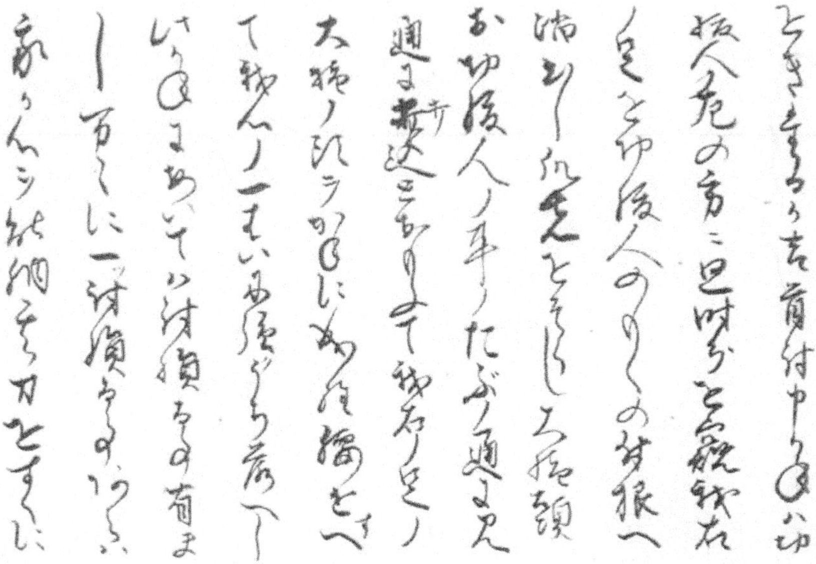

をきたるか吉、首討申かね八切
腹人ノ左の方二廻時分を窺我右
ノ足を切腹人のももの付根へ
踏出し爪先をそらし大指頭
出切腹人ノ耳ノたぶノ通に見
通に打込とおもうて我右ノ足ノ
大指ノ頭ヲ手に成程腰をすへ
て我心ノ一はいに強クうち落へし
此かねにあいて八討損る事有ま
じ、万々に一ツ打損ルコトあらバ

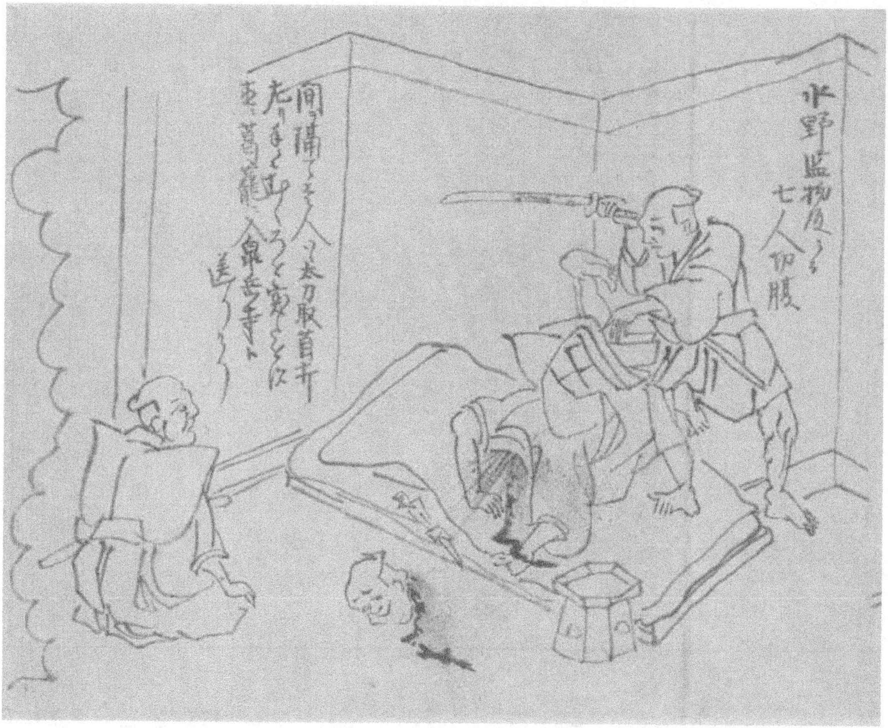

Beforehand, draw your sword in a manner that won't allow the man sentenced to Seppuku to will neither see nor hear you. Then position yourself in the spot that seems most advantageous.

○ One should strike at the moment when (the person) reaches to take the wakizashi. Also, if the person appears to be in a suitable (composed/appropriate) state

Cutting off the Head

When cutting the head, look for the moment to shift around to the left of the condemned. When cutting, step forward with your right foot, planting it beside the condemned's thigh, just where it meets the hip. Strike by lowering your hips, placing your weight on your big toe and, while aiming for the spot just below the condemned's earlobe, cut drawing on all the strength in your body. If you coordinate all these movements, you will not suffer from a miscut.

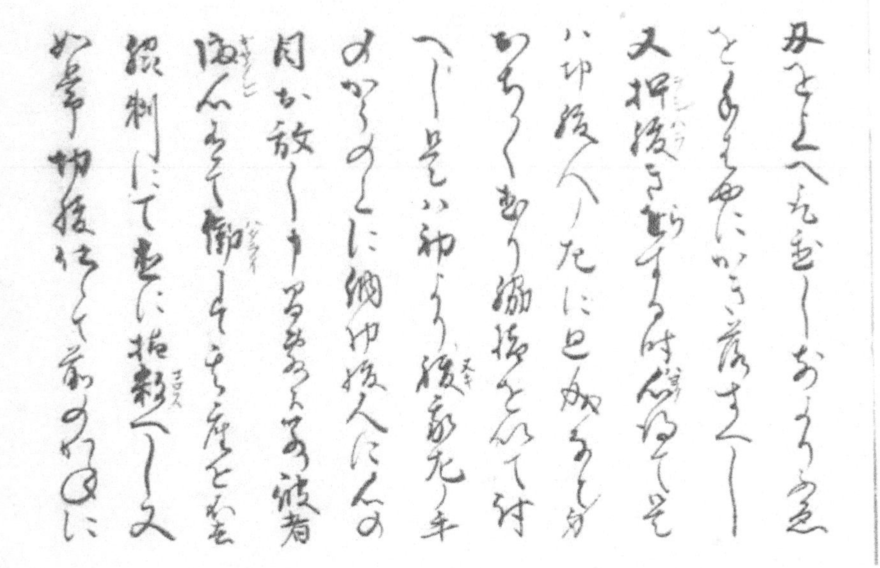

刀を上へ取置し前よりふえ
を手はやにかき落すべし
又押腹きらする時、心得て是
八切腹人ノ左に廻成など身
おちかく直り脇指を以て討
べし、是八初より抜我左ノ手
のこうの上に納切腹人に心の
目お放し申間敷候若彼者
逆心有て働にて其座を不去
脇刺にて直に指殺へし、又
如常腹切仕候て前のかねに

In the unlikely event your cut goes awry, instead of bringing your sword up again and then cutting, reapply pressure, forcing your blade down, continuing your original cut.

In some cases, the person sentenced to Seppuku will hesitate for a long time or otherwise show unwillingness to proceed. This means the official in charge of the ceremony, seated to the condemned's right must speak to the condemned and try to convince him to proceed. If this occurs, use that chance to rotate around to the condemned's left and cut him down with your sword.

In this situation you should already have your sword drawn. Stand with your right hand gripping the sword and your left palm on the back of your right hand. Do not allow your concentration to slip from the condemned even for a moment. If the condemned man begins to show signs that he is going to resist, immediately run him through with your sword until he is dead.

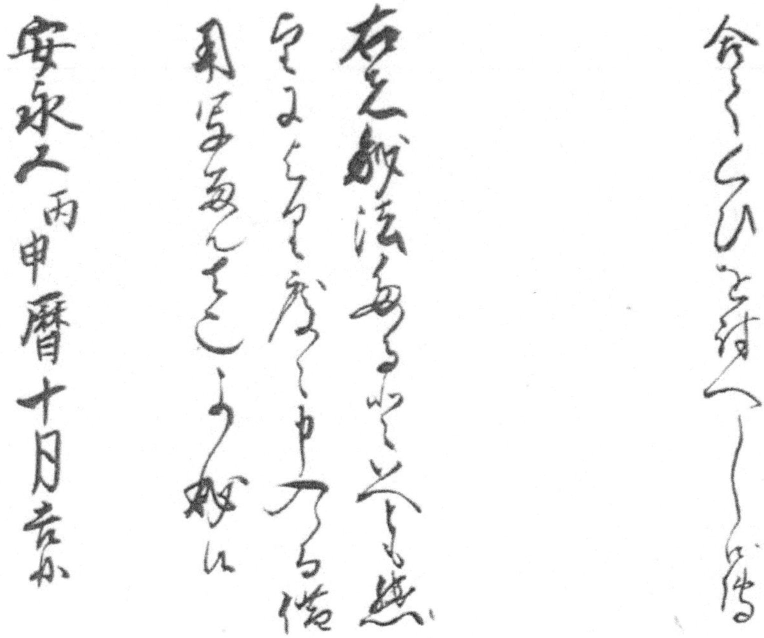

合てくびを討べし　口伝

右者秘法たるといへども懇 望によ
り度々申入らる、借 用写し留ル者
也　可秘候 安永五丙申暦十月吉日

Demonstration of how the Kaishaku's cut should look (final position of sword after cutting.)
Kendo shinzui to shidoho shosetsu 剣道神髄と指導法詳説
The Essence of Kendo and a Detailed Explanation of Teaching Methods
By Tanida Saichi 谷田左一 (?~?)1935

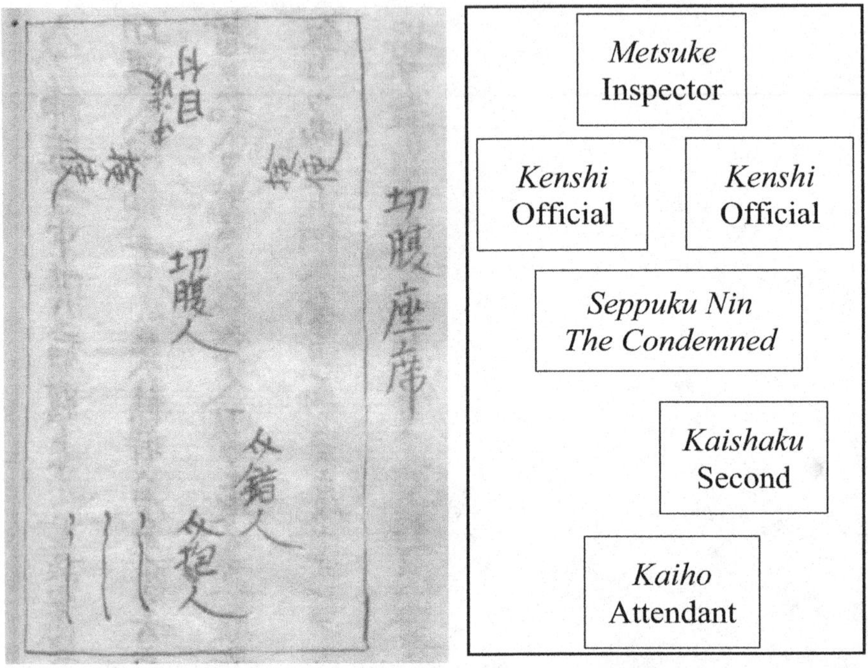

Diagram showing how a Seppuku ceremony should be set up
Secrets of Jikishin Kage School 直心影流秘伝
Kaishaku Scroll 介借書西 Edo Era

If he commits Seppuku in the proper fashion, decapitate him as previously stated. There is a Kuden, orally transmitted teaching.

Though all the methods contain in this book are secrets, due to repeated earnest requests I have agreed to copy them down. These methods should always be kept secret.

Written on an auspicious day
The tenth month of the fifth year of Anei (1776)

Secrets of Moen School Shuriken

Edo Era

Author Unknown
京都大学附属図書館 Main Library, Kyoto University

Translator's Introduction

Map of Japan with Yamaguchi Prefecture highlighted

Translator's Introduction

Moen School Shuriken was founded by Koshiba Sobe 小柴惣兵衛 and was taught primarily in Chosu Domain, modern day Yamaguchi prefecture.

A Densho owned by Fukuda Nobumasa contains the following

> Moen Ryu Shuriken originated in Tang Dynasty China and was first introduced to our nation (Japan) by Koshiba Sobe. Everyone who studied it from then on learned it based on his teachings. Koshiba taught how to throw, train this art and remember the techniques in a document containing ten entries. Many people studied this art.
>
> The Uchi Ho, way of throwing, is a *Choku-uchi Ho*, Straight Throw Method. Moen school has a unique throwing stance, whereby the user brings the Shuriken up to Miken, the spot between your eyebrows, before throwing.

Imamura Yoshio 今村嘉雄 (1903~1997)
日本武道大系 第7巻
1982

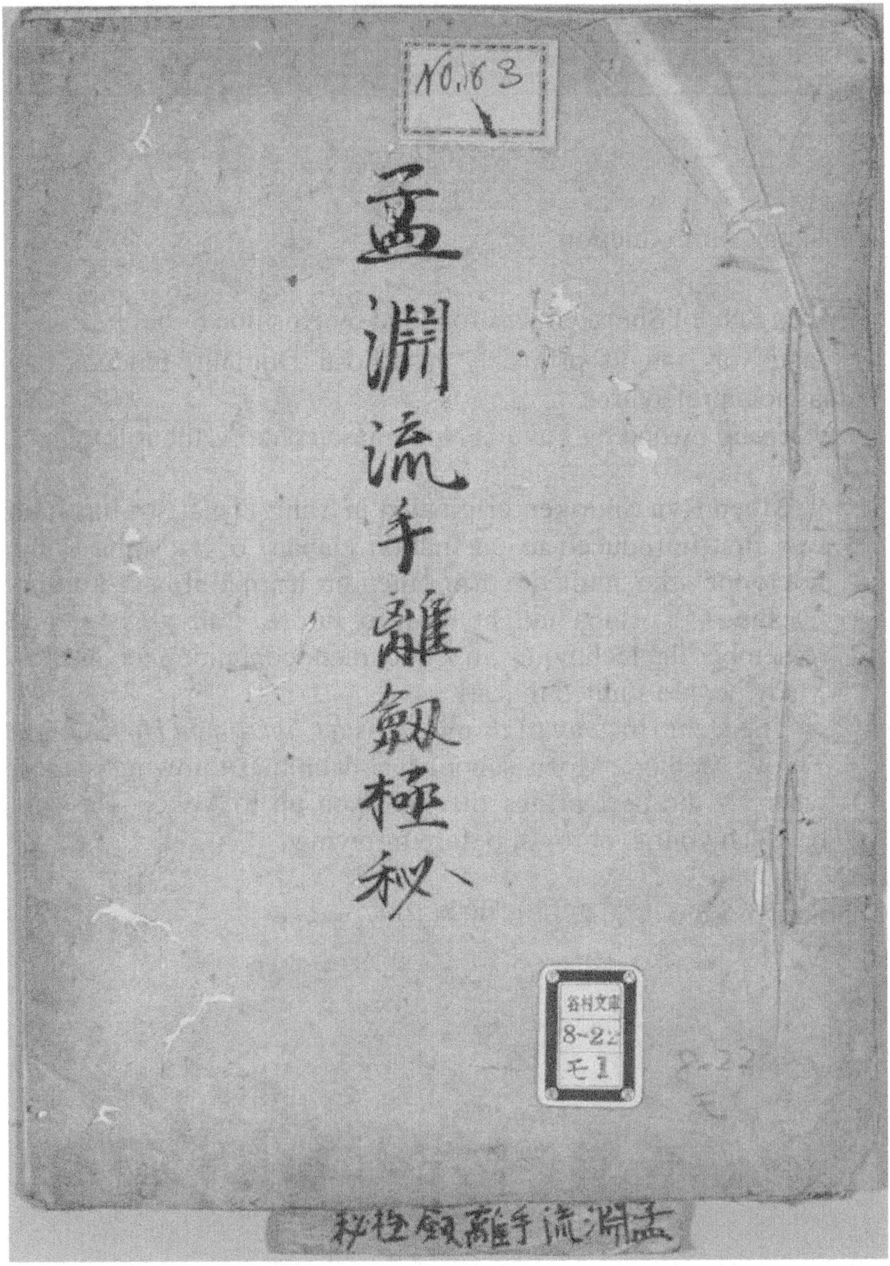
孟淵流手離劔極秘
谷村文庫
8-22
モ1

孟	Mo
淵	En
流	Ryu
手	Shu
離	Ri
剣	Ken
極	Goku
秘	Hi

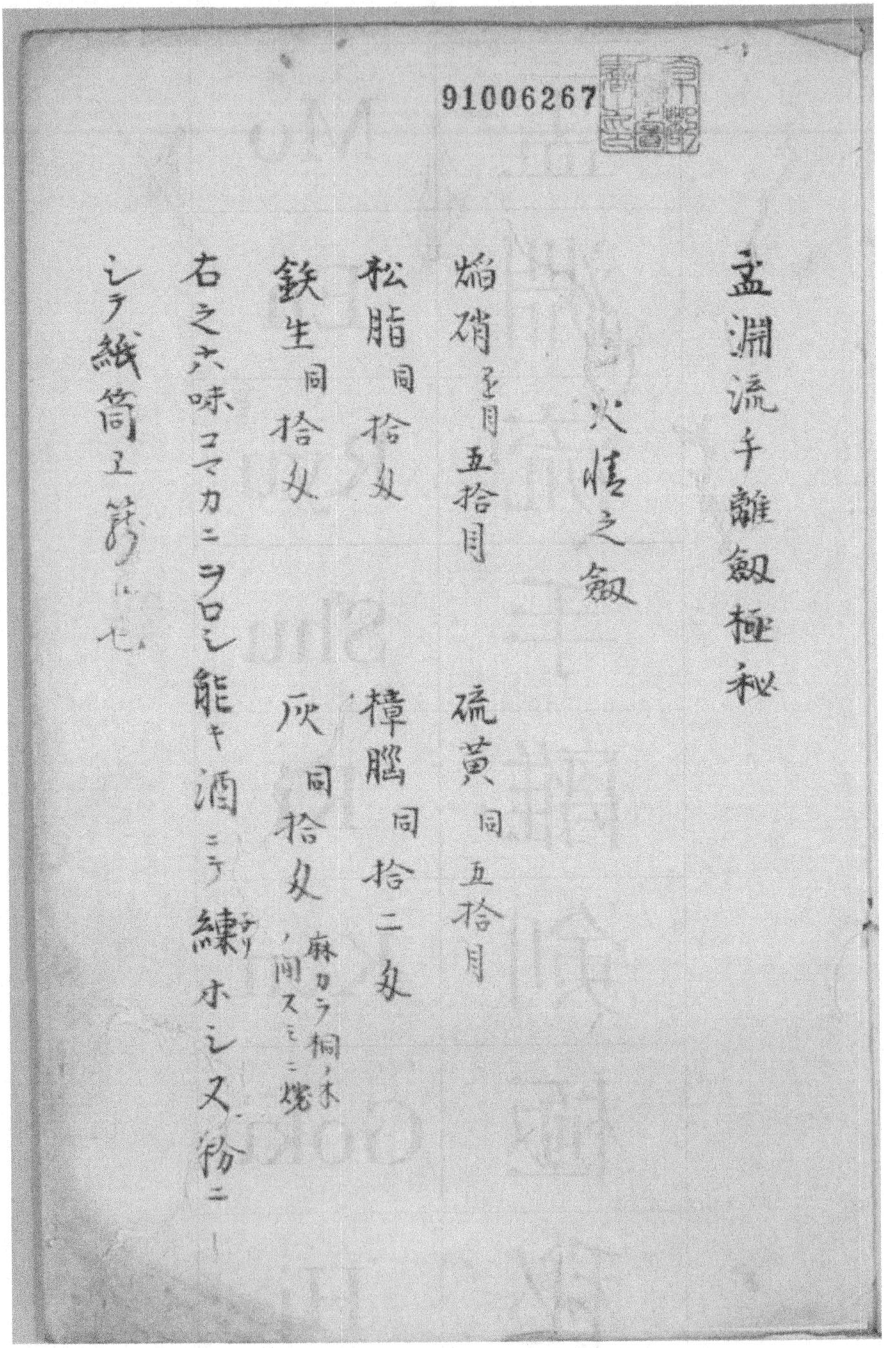

孟淵流手離剱極秘

火情之剱

焔硝壱目五拾目　硫黄同五拾目

松脂同拾匁　樟脳同拾二匁

鉄生同拾匁　灰同拾匁 麻カラ桐ノ木ノ間スミニ焼

右之六味コマカニヨロシ能キ酒ニテ練リホシ又粉ニシ

シテ紙筒エ筋に七

孟淵流手離剣極秘

火性[?]之劔
焔硝「塩硝」少目[?]　五拾目
松脂同拾匁
鉄生同拾匁
硫黄同五拾目
樟脳同拾二匁
灰同拾匁
麻カラ桐ノ木ノ間スミニ焼

右之六味コマカニヲロシ能キ酒ニテ練ホシ粉
ニ
シテ紙筒ヱ詰ル也

Secrets of Moen School Shuriken

Kasei no Ken – Sword infused with fire

Saltpeter – ground finely, 50 monme (6.6 oz)
Pine resin – same, 10 monme (1.3 oz)
Iron filings – same, 10 monme (1.3 oz)
Sulfur – same, 50 monme (6.6 oz)
Camphor – same, 12 monme (1.6 oz)
Ash – same, 10 monme (1.3 oz) Make the ash from burnt hemp and/or paulownia wood

Finely grind together the above six ingredients together with high-quality Sake. Dry the mixture and powder it and pack it into a paper tube.

道火薬之法

焔硝正目五匁　硫黄同六匁
灰同一匁　樟脳同一匁
右能粉ニシテ紙ヲ以薬ヲ包ミ筆ノ軸程ニヨリ
ノリニテカタメ筒ノ穴ヲ明ケタル所エノリヲ付テ
差込ヌケサル様ニシテ火矢ヲ射ント思時道火
ノコヨリノ先エ火ヲ付テ射ル也道火静ニ燃テ
シバラク有テ紙筒ノ内エ火移レハ大キニモエテ

道火矢之法
塩硝三目五匁
灰同一匁
硫黄同六匁
樟脳同一匁
右能粉ニシテ紙ヲ以テ茶ヲ包ミ筆ノ軸程ニヨリ
ノリニテカタメ筒ノ穴ヲ明ケタル所ヱノリ付テ
差込ヌクサル様ニシテ火矢ヲ射ルト思時道火
ノコヨリノ先ヱ火ヲ付テ射ル也道火静ニ燃テ
シバラク有テ紙筒ノ内ヱ火移レハ大キニモヱテ
水ヲ掛テモ消カタシ

Michibi Cha no Ho
How to Make a Tea Fuse

Saltpeter – finely ground, 5 monme (0.66 oz)
Ash – same, 1 monme (0.13 oz)
Sulfur – same, 6 monme (0.79 oz)
Camphor – same, 1 monme (0.13 oz)

Grind the above well into a powder, and roll up the Cha, "tea" or in this case the mixture you made, in paper, forming a tube about the thickness of a writing brush shaft. Cover the tube in paste to harden it. Open a hole in the tube and attach a twisted fuse made of *Koyori* paper string. Use paste to keep this in place. The construction is similar to when attaching a fuse to a fire arrow, so keep that in mind

The Michibi fuse burns quietly, and after a short while, when the fire reaches the paper tube, it will begin to burn vigorously. Even if you pour water on it, it is difficult to extinguish.

水ヲ掛テモ消カタシ

茶斂

コシヨウノ粉

タデバイ

トウガラシ

右三色等分ニシテ玉子ノカラヘ入斂ヲ打様ニ

面テイヘ投付レハ茶毒目鼻ヘ入動事ナラズ

紙ニヒ子リテモ時ノ見合ニシテ用ユル

茶劔
タデバイ
トウガラシ
コレショウノ粉
右三色等分ニシテ玉子ナシカラニ入劔ヲ打様ニ
面テイヱ投付レバ茶毒目鼻ヱ入動叓ナラズ紙ニ
ヒ子リテモ時ノ見合ニシテ用ユル也

Chaken
Tea Sword

Ash of bitter smartweed
Chili pepper
Powdered black pepper

Mix the above three ingredients in equal proportions and fill an empty eggshell with the powder.

Throw it directly at your opponent as if you were trying to strike them in the face with a sword. striking them with a sword. This will cause the “tea poison” to enter your opponent’s eyes and nose, rendering him unable to move.

It may also be wrapped in paper and used according to the situation.

毒叙

グハズ

右三味ヲ合セ面テイ狛ノ内ニテモ投付其氣目鼻エ入レ死スル也毒茶随分コマカニ粉ニシテ打離シノ手前茶ノ調合大口傳ユヘ此所エ略之外見ヲハヾカル者也

毒劒

ゲ　　ハ　　スズ

右三味ヲ合セ面テイリ狗ノ内ニテモニテ投付
其気目鼻ヱ入テ死スル也毒茶随分カコカニ粉ニ
シテ打離レノ手前茶ノ調合大口傳ユヘキ所ヱ略
之外見ヲハバカル者也

Ge	**Ha**	**Suzu**[43]

Combine the above three ingredients, and place them into a dog container then throw it in your enemy's face. If the powder enters the eyes or nose, it will cause death.

The Doku Cha, poison tea, should be ground as finely as possible. The way this tea is made and the proper distance for striking are important Kuden, orally transmitted lessons, and will not be recorded here so outsiders will not learn of these secrets.

---

[43] These are code words for the three ingredients.

目無シ鏑ノ如ク紙ヲ以コレラヘ焼薬ヲカタク
矢尻ヘ括付ケ矢柄ニハ細根ノ小ヲスゲテ右ノ
鏑ノ木ニホリキロヲアケ其口ヘ道火之薬ヲ
紙ヲコウヨリニシテ粘ニテカタク差込火ヲ付テ
射ル也道火之薬ハ随分静ニモユル薬法ナリ
早クモヘ付ケハ返テ用ニ不立ユヘ道火以火矢
ノ箭ノ秘蜜トスル也

目無シ鏑ニ如ク紙ヲ以コシラヘ焼茶ヲカタク
矢尻ヱ取付ケ矢**栖「矢筈？」**ニハ羽根ノ小ヲスゲ
テ右ノ
鏑ノ末ニホリキロヲアケ其口ヱ道火之薬ヲ
紙ヲコウヨリニシテ粘ニテカタク付込火ヲ付テ
射ル也道火之薬ハ随分静ニモユル茶法ナリ
早クモヱ付ケハ返テ用ニ不立ユヱ道火之以火矢
ノ箭ノ如秘宝トスル也

This method uses a Menashi Kabura, a signal arrow without holes in the end and therefore silent. Pack the head of the arrow firmly with Yaki-Cha, fire tea, pack it firmly with burning tea (combustible powder) and attach it to the arrowhead.

Fletch the arrow with small feathers at the end. Bore a hole in the end of the Kabura arrowhead and feed a Michibi, fuse, made of Koyori paper string into the opening. Secure it firmly with paste. Then light it and shoot the arrow.

If you make a fuse with this "tea fuse" compound, this method will give you a quietly and smoothly burning fuse.

You wouldn't use this method if you want to light a fire arrow quickly. However, if you need a fire arrow with a slow-burning fuse, then this method is a secret treasure.

Translator's Note: Signal Arrow

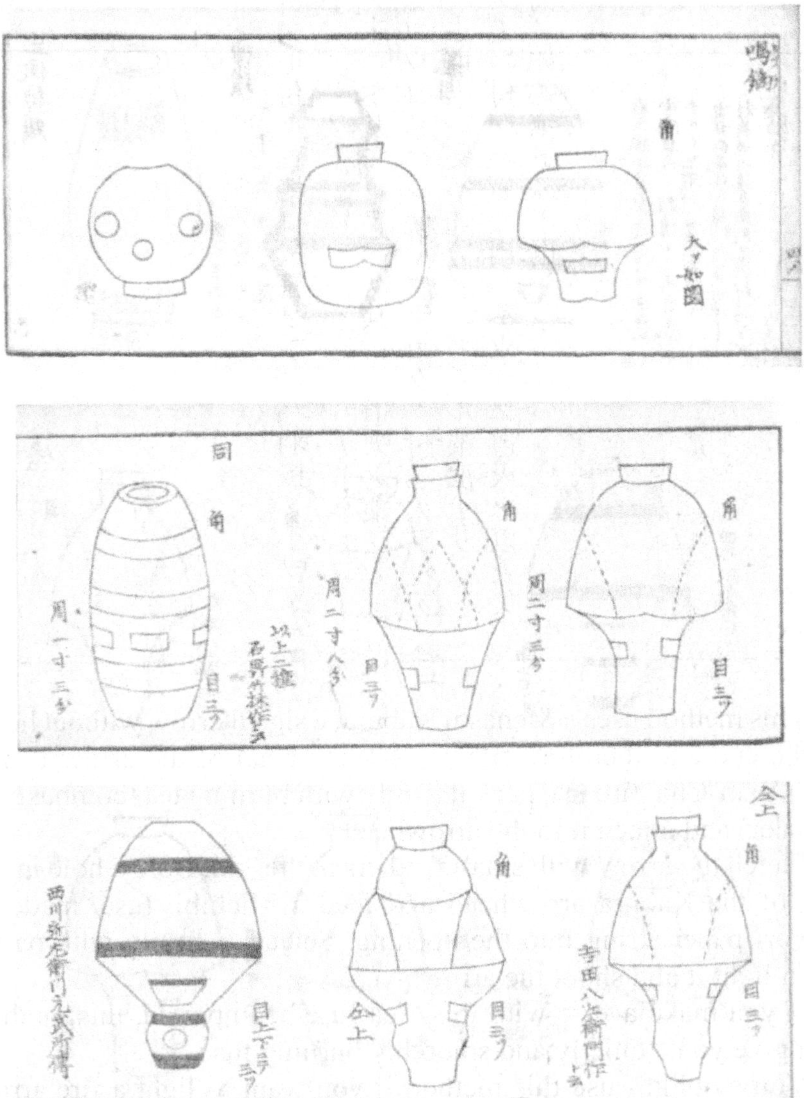

Various styles of Kabura, signaling arrowheads
From: *Diagrams of Tools of the Archer* 弓箭図式 1843
By Kurihara Nobumitsu 栗原信充

Translator's Note: Signal Arrow

According to the Abbreviated Record of Fuso 扶桑略記 (30 volumes covering 6th~11th century) in the year 940 when celebrating the surrender of Tairya no Masakado, a Kaburaya Signal Arrow was shot from the alter at Enrekiji Temple towards the east, surprising a great number of people. From that point on it became associated with misfortune befalling the enemy soldiers and began being used in ceremonies done before battle. Yoshihiko Sasama concludes that,

> The mysterious whistling sound of the Kaburaya Arrow clearly had an effect on people.

The so-called Turnip Arrows get their name from their shape. These have been discovered in ancient Japanese tombs and are thought to be related to the *Koshi* 嚆矢 "Calling Arrow" from China. In all likelihood the Koshi arrow was brought over by early settlers or travelers to Japan. The arrows are also known as *Nari-Kabura*, or Shrieking Turnip Arrows, and written with either one Kanji 鏑 or two 鳴鏑. There is a passage in the *Miscellanea of Teijo* 貞丈雑記 states,

> Before battle there are many examples of an army archer shooting a series of five Kaburaya arrows each fletched with the feathers of a different bird. The fletching should be from the Mountain Pheasant, Black Kike, Crow, Heron, Australian Bee Eater (rainbow bird).

The illustration on the following page is from Yumiya no Sho 弓矢之書 Bow and Arrow Scroll, Edo Era. This page shows several types of arrows with whistling arrowheads.

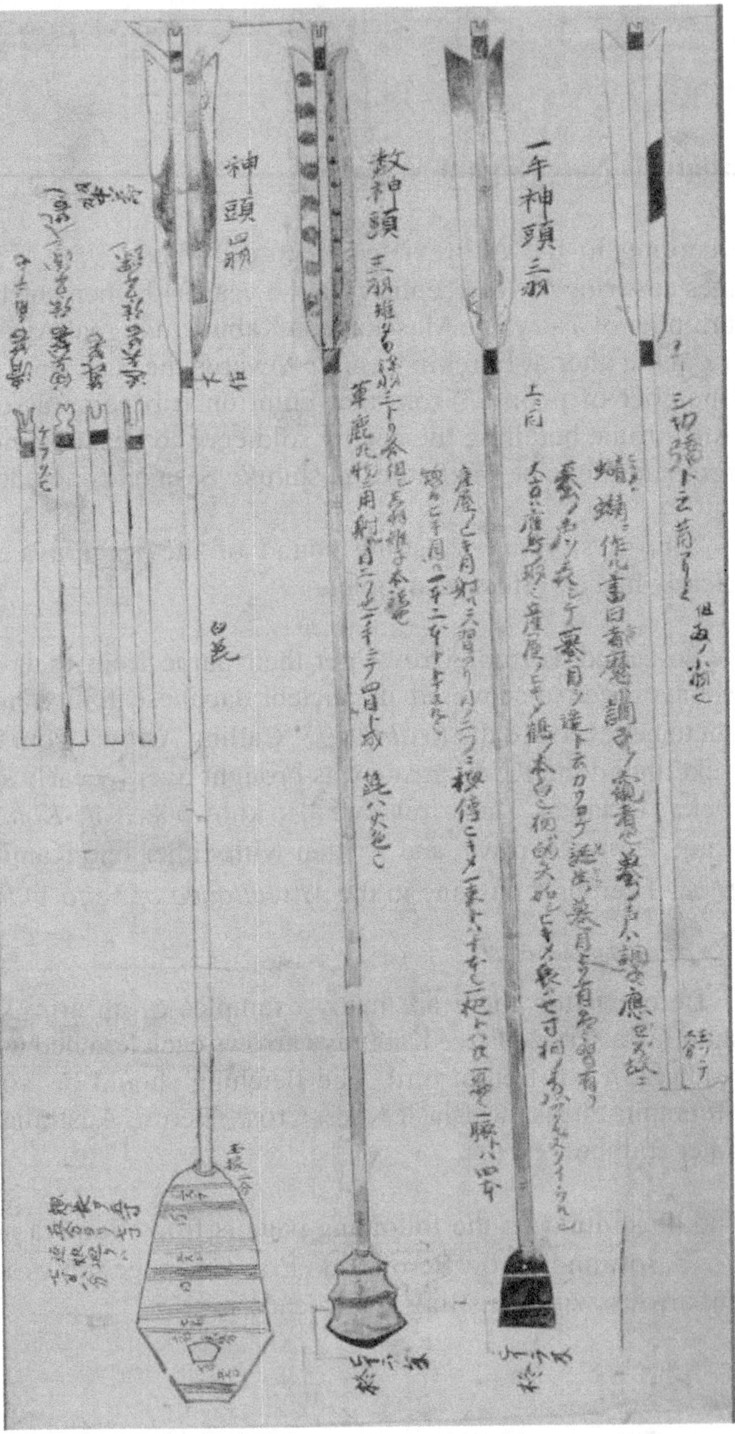
一手神頭三羽

Translator's Note: Moen School Shuriken

Fujita Seiko recorded several examples of Moen School Shuriken.

Moen School Length : 4 Sun 4 Bun 4 Ri 13.5 cm The thickest part at the cross-section is 1 Bun 5 Ri 4.5 mm	Moen School Length : 5 Sun 5 Bun 16.5 cm Diameter across square bottom : 3 Bun 9 mm	Moen School Length : From 3 to 3.5 Sun 9 ~ 10.5 cm

孟淵流

本流ではこれを宝勝剱という。

孟淵流
Moen School

In the Moen School this Shuriken is called the Treasured Victory Sword

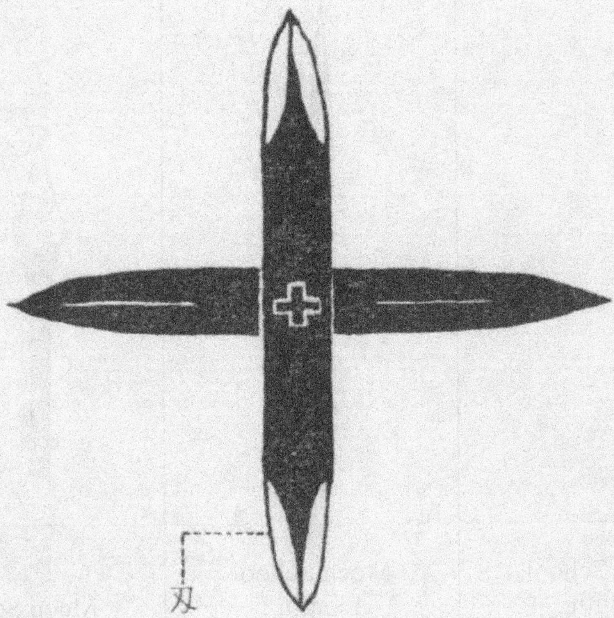

Blade

Made in the USA
Monee, IL
14 September 2026

62974504R00089